Springer Proceedings in Mathematics & Statistics

Volume 172

Springer Proceedings in Mathematics & Statistics

This book series features volumes composed of selected contributions from workshops and conferences in all areas of current research in mathematics and statistics, including operation research and optimization. In addition to an overall evaluation of the interest, scientific quality, and timeliness of each proposal at the hands of the publisher, individual contributions are all refereed to the high quality standards of leading journals in the field. Thus, this series provides the research community with well-edited, authoritative reports on developments in the most exciting areas of mathematical and statistical research today.

More information about this series at http://www.springer.com/series/10533

Valarmathi Sigamani · John J.H. Miller
Ramanujam Narasimhan
Paramasivam Mathiazhagan
Franklin Victor
Editors

Differential Equations and Numerical Analysis

Tiruchirappalli, India, January 2015

Springer

Editors
Valarmathi Sigamani
Department of Mathematics
Bishop Heber College
Tiruchirappalli, Tamil Nadu
India

Paramasivam Mathiazhagan
Department of Mathematics
Bishop Heber College
Tiruchirappalli
India

John J.H. Miller
Institute for Numerical Computation and Analysis
Dublin
Ireland

Franklin Victor
Department of Mathematics
Bishop Heber College
Tiruchirappalli
India

Ramanujam Narasimhan
Department of Mathematics
Bharathidasan University
Tiruchirappalli
India

ISSN 2194-1009 ISSN 2194-1017 (electronic)
Springer Proceedings in Mathematics & Statistics
ISBN 978-81-322-3596-5 ISBN 978-81-322-3598-9 (eBook)
DOI 10.1007/978-81-322-3598-9

Library of Congress Control Number: 2016943445

Mathematics Subject Classification (2010): 65L10, 65L11, 65L12, 65M06, 34K26, 34K28

Printed on acid-free paper

This Springer imprint is published by Springer Nature
The registered company is Springer (India) Pvt. Ltd.

Preface

This book is the consolidation of the selected articles by the participants of the International Winter Workshop on Differential Equations and Numerical Analysis (DEANA 2015), held during 5–7 January 2015 at Bishop Heber College, Tiruchirappalli, India. Though the conference was intended to accommodate works on differential equations, the main concentration was on equations whose solutions and their derivatives are non-smooth with singularities related to boundary layers. Most of the works were on singular perturbation problems whose solutions exhibit initial/interior/boundary layers and occur in many physical phenomena.

With the presentation of the paper "On the motion of fluids with very little friction" by Ludwig Prandtl in 1904, in the International Congress of Mathematics, held in Heidelberg, Germany, the field of classical fluid dynamics got revolutionized. This led to the development of boundary layer theory and singular perturbation problems. Typically, these problems arise in various fields of applied mathematics such as fluid dynamics (boundary layer problems), elasticity (edge effect in shells), quantum mechanics (WKB problems), electrical networks, chemical reactions, control theory, gas porous electrodes theory and many other areas. The Navier–Stokes' equation with a large Reynolds number is one of the most striking examples of singular perturbation problems.

The aim of the conference was to give the young researchers the core of the subject for which pioneers in this area of research were invited from India and abroad. The invited talks were given by Prof. John J.H. Miller, Professor Emeritus, Trinity College, Dublin and Director, INCA, Dublin, Ireland; Prof. Eugene O'Riordan, School of Mathematical Sciences, Dublin City University, Ireland; Prof. N. Ramanujam, Honorary Professor, Department of Mathematics, Bharathidasan University, Tiruchirappalli, India; Dr. S. Valarmathi, Associate Professor and Head, Department of Mathematics, Bishop Heber College, Tiruchirappalli, India; and a few others. Also, there were contributions from various researchers working on differential equations and numerical analysis.

The book consists of two parts. Part I includes lectures by the invited speakers. The chapter "Elementary Tutorial on Numerical Methods for Singular Perturbation

Problems" gives a tutorial on singular perturbation problems. The chapter entitled "Interior Layers in Singularly Perturbed Problems" presents an introduction to interior layers occurring in the solution of singular perturbation problems. In the chapter "Singularly Perturbed Delay Differential Equations and Numerical Methods", an introduction about the applications and various methods of solving delay differential equations are presented. In the chapter "Initial or Boundary Value Problems for Systems of Singularly Perturbed Differential Equations and Their Solution Profile", a sketch of the analytical and numerical results for initial/boundary value problems for systems of singularly perturbed differential equations is given. In Part II, six refereed contributions of people working in the area of singular perturbation problems are included.

We are grateful to the invited speakers, the authors of contributed papers and to the unnamed referees for their valuable contributions, without which this volume is not possible. We acknowledge with sincere thanks the financial support extended by the University Grants Commission, Government of India and the National Board for Higher Mathematics, Government of India, to conduct the conference. Thanks are also due to the members of the organizing committee and the Principal and the management of Bishop Heber College. Our special thanks are also due to Mr. Kennet Jacob Jaisingh, software consultant, who designed the logo of DEANA 2015, constructed the website and helped us to have the book of abstracts, the brochure, etc., in the stipulated time.

Tiruchirappalli, India — Valarmathi Sigamani
Dublin, Ireland — John J.H. Miller
Tiruchirappalli, India — Ramanujam Narasimhan
Tiruchirappalli, India — Paramasivam Mathiazhagan
Tiruchirappalli, India — Franklin Victor
April 2016

Contents

Editors and Contributors

About the Editors

Valarmathi Sigamani is an Associate Professor & Head at the Department of Mathematics at Bishop Heber College, Tiruchirappalli, Tamil Nadu, India. She did her doctoral studies in numerical analysis at Bharathidasan University, Tiruchirappalli, under the guidance of Prof. N. Ramanujam. She has been teaching algebra, analysis, topology, calculus of variations, computer programming and differential equations to undergraduate and postgraduate students of mathematics. She has published 20 research articles in various reputed international journals and has guided 5 students and guiding some more in their doctoral studies in singular perturbation problems and numerical analysis. Currently, she is working on the qualitative and quantitative analysis of singularly perturbed multi-scale dynamical systems and their applications in biology. She is a council member of the Institute of Numerical Computation and Analysis (INCA), Dublin, Ireland, and is a member in various mathematical societies.

John J.H. Miller is a director of the Institute for Numerical Computation and Analysis (INCA), Dublin, and Professor Emeritus of Mathematics at the Trinity College, Dublin. He obtained two undergraduate degrees at Trinity College, University of Dublin, Ireland: one in mathematics and the other in natural sciences. He then studied numerical analysis at the Massachusetts Institute of Technology under the supervision of Gilbert Strang, where he was awarded a Ph.D. degree. His mathematical interests include power bounded matrices, the zeros of polynomials, numerical methods for modelling semiconductor devices, asymptotic and computational methods for problems exhibiting boundary and interior layers, data analysis of signals from neurons and, recently, the mathematical modelling and construction of oscillating water column devices for generating electric power from ocean waves. He has written over 100 papers on these topics, as well as three research monographs.

Ramanujam Narasimhan is an Honorary Professor of Mathematics at Bharathidasan University, Tiruchirappalli, Tamil Nadu. He worked under the guidance of Dr. U.N. Srivastava and got his Ph.D. degree from the Indian Institute of Technology, Madras. His research interests include differential equations and numerical analysis. He has published over 85 papers in various international journals and has guided 11 students in their doctoral studies. He has been teaching various courses such as ordinary differential equations, partial differential equations, numerical analysis, functional analysis, fluid dynamics, computer programming, real analysis, integral equations and calculus of variations. He has served in many responsible positions including the vice chancellor (in-charge) of Bharathidasan University.

Paramasivam Mathiazhagan is an Assistant Professor of Mathematics at Bishop Heber College, Tiruchirappalli, Tamil Nadu. He got his Ph.D. from Bharathidasan University in 2013 under the guidance of Dr. S. Valarmathi. His research interests include numerical analysis and boundary layer theory. He has worked on the multi-scale reaction–diffusion problems and has published articles in reputed international journals. He has been teaching various courses such as partial differential equations, functional analysis, fluid dynamics and computer programming. Apart from mathematics, he is a brilliant cricketer who played for the university during his college days.

Franklin Victor is an Assistant Professor of Mathematics at Bishop Heber College, Tiruchirappalli, Tamil Nadu. He completed his doctoral studies in numerical methods for solving singularly perturbed partial differential equations of parabolic type under the guidance of Dr. S. Valarmathi. Recently, he got interested in delay differential equations and is currently working on numerical methods to solve singularly perturbed delay differential equations. His other research interests are mathematics of finance and Black–Scholes equations. He has been teaching various courses such as mathematical analysis, ordinary differential equations and partial differential equations. Besides mathematics, he is a talented musician. He is a reputed choral arranger and a conductor of various ensembles of his region.

Contributors

P. Avudai Selvi Department of Mathematics, School of Mathematical Sciences, Bharathidasan University, Tiruchirappalli, Tamil Nadu, India

Franklin Victor Department of Mathematics, Bishop Heber College, Tiruchirappalli, Tamil Nadu, India

Mariappan Manikandan Department of Mathematics, Bishop Heber College, Tiruchirappalli, Tamil Nadu, India

John J.H. Miller INCA-Institute for Numerical Computation and Analysis, Dublin, Ireland; Trinity College, Dublin, Ireland

Nagarajan Shivaranjani Bishop Heber College, Tiruchirappalli, Tamil Nadu, India

Eugene O'Riordan Dublin City University, Dublin 9, Ireland

Parthiban Swaminathan Department of Mathematics, Bishop Heber College, Tiruchirappalli, Tamil Nadu, India

Ramanujam Narasimhan Department of Mathematics, School of Mathematical Sciences, Bharathidasan University, Tiruchirappalli, Tamil Nadu, India

Valarmathi Sigamani Department of Mathematics, Bishop Heber College, Tiruchirappalli, Tamil Nadu, India

Part I
Invited Papers

Elementary Tutorial on Numerical Methods for Singular Perturbation Problems

John J.H. Miller

Abstract In the first section we introduce a simple singularly perturbed initial value problem for a first order linear differential equation. We construct the backward Euler finite difference method for this problem. We then discuss continuous and discrete maximum principles for the associated continuous and discrete operators and we conclude the section by defining what is meant by a parameter-uniform numerical method. In the second section we introduce a fitted operator method on a uniform mesh for our simple initial value problem defined in the previous section. We then prove rigorously that this method is parameter-uniform at the mesh points. Fitted mesh methods on piecewise uniform meshes are introduced in the third section. A fitted mesh method for our simple initial value problem is constructed. It is proved rigorously that this method is parameter-uniform at the mesh points. Finally, in the fourth section, numerical solutions of singular perturbation problems are discussed. Computations using standard and a parameter-uniform numerical method are presented. The usefulness and reliability of parameter-uniform methods is demonstrated.

Keywords Singular perturbation problems · Finite difference scheme · Shishkin mesh · Boundary layers · Parameter uniform convergence

1 Introduction to Singular Perturbation Problems and Their Numerical Solution

We begin with a brief introduction to singular perturbation problems for differential equations by describing the concepts in terms of a simple linear problem and a maximum principle.

We consider the simple singularly perturbed first order initial value problem (P_ε) on the interval $\Omega = (0, T]$

J.J.H. Miller (✉)
INCA-Institute for Numerical Computation and Analysis, Dublin, Ireland
e-mail: jm@incaireland.org

V. Sigamani et al. (eds.), *Differential Equations and Numerical Analysis*, Springer Proceedings in Mathematics & Statistics 172,
DOI 10.1007/978-81-322-3598-9_1

$$\varepsilon u'_\varepsilon + a(t)u_\varepsilon = f(t),\ t \in \Omega, \tag{1}$$

$$u_\varepsilon(0) \text{ given}, \tag{2}$$

where, for all $t \in \Omega,\ a(t) \geq \alpha$ and $0 < \varepsilon \leq 1$.

We are interested in designing a numerical method which gives good approximations to the solution of (P_ε), regardless of the value of ε, in the entire range $0 < \varepsilon \leq 1$. It can be seen from the exact solution of this problem that, as $\varepsilon \to 0$, the gradient of the solution becomes increasingly steep as we approach $t = 0$. We say that the solution has a *layer* at the point $t = 0$. In this case it is called an initial layer, because it is associated with a boundary point where an initial condition is specified. In the analysis that follows it becomes clear that the width of this layer is $O(\varepsilon)$.

To analyse such problems, and their numerical solutions, we introduce some norms and semi-norms. For singular perturbation problems it is important to work in the maximum norm. We define the maximum norm of a differentiable function on a set S by

Definition 1.1 $|\phi|_S = \sup_{t\in S} |\phi(t)|$

and, for any positive integer k, we define the kth order semi-norm of a differentiable function on a set S by

Definition 1.2 $|\phi|_{k,S} = \sup_{t\in S} |\phi^{(k)}(t)|$.

For $k = 0$, the semi-norm becomes the maximum norm. Note that when the meaning is clear the subscript S is usually dropped from the notation.
For some $T > 0$, let $\Omega = (0, T]$ and $\overline{\Omega} = [0, T]$.
For convenience we introduce the differential operator

$$L_\varepsilon = \varepsilon \frac{d}{dt} + a(t),\quad a(t) > \alpha > 0,\quad \text{for all}\ \ t \in \overline{\Omega}.$$

The operator L_ε satisfies the following maximum principle.

Lemma 1.1 *Let $\psi(t)$ be any function in the domain of L_ε such that $\psi(0) \geq 0$. Then $L_\varepsilon\psi(t) \geq 0$ for all $t \in \Omega$ implies that $\psi(t) \geq 0$ for all $t \in \overline{\Omega}$.*

Proof Let t^* be such that $\psi(t^*) = \min_t \psi(t)$ and assume that the lemma is false. Then $\psi(t^*) < 0$. From the hypotheses we have $t^* \neq 0$ and $\psi'(t^*) \leq 0$. Thus

$$L_\varepsilon\psi(t^*) = \varepsilon\psi'(t^*) + a(t^*)\psi(t^*) < 0,$$

which contradicts the assumption.

This leads immediately to the following stability result.

Lemma 1.2 *Let $\psi(t)$ be any function in the domain of L_ε. Then*

$$|\psi(t)| \leq \max\{|\psi(0)|, \frac{1}{\alpha}|L_\varepsilon\psi|\}\ \text{ for all }\ t \in \overline{\Omega}.$$

Proof Define the two functions

$$\theta^{\pm}(t) = \max\{|\psi(0)|, \frac{1}{\alpha}|L_\varepsilon\psi|\} \pm \psi(t).$$

It is not hard to verify that $\theta^{\pm}(0) \geq 0$ and $L_\varepsilon\theta^{\pm}(t) \geq 0$. It follows from Lemma 1.1 that $\theta^{\pm}(t) \geq 0$, for all $t \in \overline{\Omega}$, as required.

In order to discuss numerical solutions we need to discretise the domain $\Omega = (0, T]$. The simplest discretisation is a uniform mesh having N sub-intervals of equal length h, which is determined by a set of $N+1$ equally spaced mesh points $\Omega_h^N = \{t_i\}_{i=0}^N$. Here, $t_0 = 0, \quad t_N = T$, and, for any $i, \; 1 \leq i \leq N, \quad h = t_i - t_{i-1}$.

Uniform meshes are adequate for fitted operator methods, but non-uniform meshes are used for fitted mesh methods. Piecewise-uniform meshes are the simplest kind of non-uniform mesh. We will be interested in piecewise-uniform meshes, where the transition point between the two uniform meshes is specially chosen. This choice depends on the problem under consideration.

We now consider discrete solutions. We first need to introduce the forward, backward, centered and second order finite difference operators $D^+ \; D^- \; D^0 \quad \delta^2$, where, on an arbitrary mesh $\Omega^N = \{t_i\}_{i=0}^N$,

$$D^+U(t_j) = \frac{U(t_{j+1}) - U(t_j)}{t_{j+1} - t_j}, \quad D^-U(t_j) = \frac{U(t_j) - U(t_{j-1})}{t_j - t_{j-1}}$$

$$D^0U(t_j) = \frac{U(t_{j+1}) - U(t_{j-1})}{t_{j+1} - t_{j-1}}, \quad \delta^2U(t_j) = \frac{D^+U(t_j) - D^-U(t_j)}{(t_{j+1} - t_{j-1})/2}.$$

Using these we can now define our finite difference methods. For example, the backward Euler finite difference method for (P_ε) is the following finite difference method (P_ε^N)

$$\varepsilon D^-U^N + a(t)U^N = f, \quad U^N(0) = u(0),$$

or in operator form

$$L_\varepsilon^N U^N = f, \quad U^N(0) = u(0),$$

where the finite difference operator is defined by

$$L_\varepsilon^N = \varepsilon D^- + a(t)I.$$

We have the following discrete maximum principle for L_ε^N analogous to the continuous case.

Lemma 1.3 *For any mesh function* Ψ^N, *the inequalities* $\Psi^N(0) \geq 0$ and $L_\varepsilon{}^N\Psi^N(t_j) \geq 0$ *for* $1 \leq j \leq N$, *imply that* $\Psi^N(t_j) \geq 0$ *for all* $j, \;\; 0 \leq j \leq N$.

Proof Let j^* be such that $\Psi^N(t_{j^*}) = \min_j \Psi^N(t_j)$ and assume that the lemma is false. Then $\Psi^N(t_{j^*}) < 0$. From the hypotheses we have $j^* \neq 0$ and $\Psi^N(t_{j^*}) - \Psi^N(t_{j^*-1}) \leq 0$. Thus

$$L_\varepsilon^N \Psi^N(t_{j^*}) = \varepsilon \frac{\Psi^N(t_{j^*}) - \Psi^N(t_{j^*-1})}{\delta_{j^*}} + a(t_{j^*})\Psi^N(t_{j^*}) \leq a(t_{j^*})\Psi^N(t_{j^*}) < 0,$$

which contradicts the assumption, as required.

An immediate consequence of this is the following discrete stability result analogous to the continuous result.

Lemma 1.4 *For any mesh function Ψ^N, we have*

$$|\Psi^N(t_j)| \leq \max\{|\Psi^N(0)|, \frac{1}{\alpha}|L_\varepsilon^N \Psi^N|\}, \quad 0 \leq j \leq N.$$

Proof Define the two mesh functions

$$\Theta_\pm^N(t) = \max\{|\Psi^N(0)|, \frac{1}{\alpha}|L_\varepsilon^N \Psi^N|\} \pm \Psi^N(t).$$

It is not hard to verify that $\Theta_\pm^N(0) \geq 0$ and $L_\varepsilon^N \Theta_\pm^N(t_j) \geq 0$. It follows from Lemma 1.3 that $\Theta_\pm^N(t_j) \geq 0$ for all $0 \leq j \leq N$.

We now consider estimates of the local truncation error, which will be required later in our proofs of parameter-uniform convergence. In particular, we give two distinct estimates of the local truncation error in approximating $\frac{d}{dt}$ by D^-, that is we consider $|(D^- - \frac{d}{dt})\phi|$ for any differentiable function ϕ.

First, we observe that, at any mesh point t_i,

$$(D^- - \frac{d}{dt})\phi(t_i) = \frac{\phi(t_i) - \phi(t_{i-1})}{t_i - t_{i-1}} - \phi'(t_i) = \frac{1}{t_i - t_{i-1}} \int_{t_{i-1}}^{t_i} (\phi'(s) - \phi'(t_i))ds$$

and so

$$|(D^- - \frac{d}{dt})\phi(t_i)| \leq \frac{1}{t_i - t_{i-1}} \int_{t_{i-1}}^{t_i} (|\phi'(s)| + |\phi'(t_i)|)ds \leq 2|\phi|_{1,[t_{i-1},t_i]}. \quad (3)$$

Secondly, integrating by parts, we obtain

$$(D^- - \frac{d}{dt})\phi(t_i) = \frac{\phi(t_i) - \phi(t_{i-1})}{t_i - t_{i-1}} - \phi'(t_i) = \frac{1}{t_i - t_{i-1}} \int_{t_{i-1}}^{t_i} (t_{i-1} - s)\phi''(s)ds$$

and it follows that

$$|(D^- - \frac{d}{dt})\phi(t_i)| \le \frac{1}{t_i - t_{i-1}} \int_{t_{i-1}}^{t_i} (s - t_{i-1})|\phi''(s)|ds \le \frac{1}{2}(t_i - t_{i-1})|\phi|_{2,[t_i,t_{i-1}]}. \tag{4}$$

We now define what is meant by a parameter-uniform numerical method for a family of singular perturbation problems.
It is important to note that in this definition, and throughout the remainder of this tutorial, C denotes a generic constant that is independent of the singular perturbation parameter ε.

Definition 1.3 Consider a family of problems (P_ε) parameterised by the singular perturbation parameter ε, $0 < \varepsilon \le 1$. Suppose that the exact solution u_ε is approximated by the sequence of numerical solutions $\{U_\varepsilon^N\}_{N=1}^\infty$, defined on meshes Ω_ε^N, where N is the discretization parameter. Then, the numerical solutions U_ε^N are said to converge ε-uniformly to the exact solution u_ε, if there exists a positive integer N_0, and positive numbers C and p, all independent of N and ε, such that, for all $N \ge N_0$,

$$\sup_{0<\varepsilon\le 1} |U_\varepsilon^N - u_\varepsilon|_{\Omega_\varepsilon^N} \le CN^{-p}, \tag{5}$$

where $|.|_{\Omega_\varepsilon^N}$ is the maximum norm on Ω_ε^N.
We now derive two useful consequences of the stability of L_ε established in Lemma 1.2. The first is an a priori bound on the solution of (P_ε). It is an immediate corollary to Lemma 1.2.

Corollary 1.1 *Let u_ε be the solution of (P_ε). Then,*

$$|u_\varepsilon| \le |u_\varepsilon(0)| + \frac{1}{\alpha}|f|.$$

The second establishes the uniqueness of the solution.

Corollary 1.2 *If (P_ε) has a solution, it is unique.*

Proof Let u_1, u_2 be any two solutions of (P_ε) and consider $z = u_1 - u_2$. Then it is easy to see that

$$z(0) = z_1(0) - z_2(0) = 0 \text{ and}$$
$$L_\varepsilon z = L_\varepsilon(u_1 - u_2) = L_\varepsilon u_1 - L_\varepsilon u_2 = f - f = 0.$$

Hence z is a solution of (P_ε) in the special case with homogeneous initial condition and homogeneous differential equation. Applying Corollary 1.1 in this case gives, for all $t \in \overline{\Omega}$,

$$|z(t)| \le |z(0)| + \frac{1}{\alpha}|L_\varepsilon z| \le 0.$$

We conclude that $|u_1(t) - u_2(t)| = 0$, for all $t \in \overline{\Omega}$, and so $u_1 = u_2$.
The next lemma provides additional classical a priori bounds on the solution of the problem (P_ε) and its derivatives.

Lemma 1.5 *Let u_ε be the solution of (P_ε). Then, we have*

$$|u_\varepsilon|_k \leq C\varepsilon^{-k}, \text{ for each } k, \quad 0 \leq k \leq 2.$$

Proof For $k = 0$ the required result is simply Corollary 1.1. For $k = 1$ we use the differential equation as follows. We have

$$u'_\varepsilon(t) = \varepsilon^{-1}(f(t) - a(t)u_\varepsilon(t))$$

and so

$$|u'_\varepsilon| \leq \varepsilon^{-1}(|f| + |a||u_\varepsilon|) \leq C\varepsilon^{-1}$$

as required. The result for $k = 2$ follows from this result and the differential equation again. We have

$$u''_\varepsilon(t) = \varepsilon^{-1}(f'(t) - a(t)u'_\varepsilon(t) - a'(t)u_\varepsilon(t))$$

and so

$$|u''|_\varepsilon \leq \varepsilon^{-1}(|f'| + |a||u'_\varepsilon| + |a'||u_\varepsilon|) \leq C\varepsilon^{-1}(1 + \varepsilon^{-1} + 1) \leq C\varepsilon^{-2},$$

which completes the proof.
There are discrete analogues to the above. For example, the following analogue to Corollary 1.1 is an immediate consequence of Lemma 1.4.

Corollary 1.3 *Let U_ε^N be the solution of (P_ε^N). Then,*

$$|U_\varepsilon^N| \leq |U_\varepsilon^N(0)| + \frac{1}{\alpha}|f|.$$

2 Fitted Operator Methods

In this section a simple result from [1] is proved. The aim is to explain the proof as clearly as possible by including some helpful details that were omitted from the book. Consider again the initial value problem (P_ε)

$$L_\varepsilon u_\varepsilon(t) \equiv \varepsilon u'_\varepsilon(t) + a(t)u_\varepsilon(t) = f(t), \qquad u_\varepsilon(0) \text{ given.}$$

We introduce the fitted operator method (P_ε^h)

$$L_\varepsilon^h U_\varepsilon^N(t_i) \equiv \varepsilon\sigma_i(\rho)D^+U_\varepsilon^N(t_i) + a(t_i)U_\varepsilon^N(t_i) = f(t_i), \quad t_i \in \Omega^N \qquad u_\varepsilon^h(0) = u_\varepsilon(0),$$

where D^+ is the forward difference operator on the uniform mesh $\overline{\Omega}^N = \{t_i\}_0^N$, $t_i = ih,\ h = \frac{1}{N}$ and σ_i is the fitting factor

$$\sigma_i(\rho) = \frac{\rho a(ih)}{1 - e^{-\rho a(ih)}},\quad \rho = \frac{h}{\varepsilon}.$$

It is convenient to write

$$\mu_i^h = \frac{\sigma_i(\rho)}{\rho} = \frac{a(ih)}{1 - e^{-\rho a(ih)}}.$$

We consider the problem on the two distinct uniform meshes $\Omega^N = \{t_i^h\}_{i=0}^N$ and $\Omega^{2N} = \{t_{2i}^{\frac{h}{2}}\}_{i=0}^{2N}$, where $t_i^h = ih$, $U_\varepsilon^N(t_i^h) = u_i^h$ and we note that $t_i^h = ih = (2i)(\frac{h}{2}) = t_{2i}^{\frac{h}{2}}$.
The finite difference operator can now be written in the form

$$L_\varepsilon^h u_i^h = \mu_i^h(u_{i+1}^h - e^{-\rho a(ih)} u_i^h).$$

We now estimate

$$L_\varepsilon^h(u_{2i}^{\frac{h}{2}} - u_i^h). \tag{6}$$

Considering separately the two terms in (6) we have

$$L_\varepsilon^h u_i^h = f(ih) \tag{7}$$

and

$$L_\varepsilon^h u_{2i}^{\frac{h}{2}} = \mu_i^h(u_{2i+2}^{\frac{h}{2}} - e^{-\rho a(ih)} u_{2i}^{\frac{h}{2}}). \tag{8}$$

Now express $u_{2i+2}^{\frac{h}{2}}$ in terms of $u_{2i}^{\frac{h}{2}}$ using the following two equations

$$L_\varepsilon^{\frac{h}{2}} u_{2i}^{\frac{h}{2}} = \mu_{2i}^{\frac{h}{2}}(u_{2i+1}^{\frac{h}{2}} - e^{-\frac{\rho}{2}a(ih)} u_{2i}^{\frac{h}{2}}) = f(ih),$$
$$L_\varepsilon^{\frac{h}{2}} u_{2i+1}^{\frac{h}{2}} = \mu_{2i+1}^{\frac{h}{2}}(u_{2i+2}^{\frac{h}{2}} - e^{-\frac{\rho}{2}a((2i+1)\frac{h}{2})} u_{2i+1}^{\frac{h}{2}}) = f((2i+1)\frac{h}{2}).$$

To eliminate $u_{2i+1}^{\frac{h}{2}}$ between these two equations, first divide each equation by the appropriate μ, then multiply the first equation by $e^{-\frac{\rho}{2}a((2i+1)\frac{h}{2})}$ and add to get

$$u_{2i+2}^{\frac{h}{2}} = e^{-\frac{\rho}{2}(a((2i+1)\frac{h}{2})+a(ih))} u_{2i}^{\frac{h}{2}} + e^{-\frac{\rho}{2}a((2i+1)\frac{h}{2})} \frac{f(ih)}{\mu_{2i}^{\frac{h}{2}}} + \frac{f((2i+1)\frac{h}{2})}{\mu_{2i+1}^{\frac{h}{2}}}.$$

Using this expression for $u_{2i+2}^{\frac{h}{2}}$, the right hand side of (8) becomes

$$\mu_i^h(e^{-\frac{\rho}{2}(a((2i+1)\frac{h}{2})+a(ih))} - e^{-\rho a(ih)})u_{2i}^{\frac{h}{2}} +$$
$$\frac{\mu_i^h}{\mu_{2i}^{\frac{h}{2}}}e^{-\frac{\rho}{2}a((2i+1)\frac{h}{2})}f(ih) + \frac{\mu_i^h}{\mu_{2i+1}^{\frac{h}{2}}}f((2i+1)\frac{h}{2}).$$

Using the Taylor expansion

$$f((2i+1)\frac{h}{2}) = f(ih) + \frac{h}{2}f'(\xi), \quad ih \le \xi \le (2i+1)\frac{h}{2}, \tag{9}$$

this becomes

$$\mu_i^h(e^{-\frac{\rho}{2}(a((2i+1)\frac{h}{2})+a(ih))} - e^{-\rho a(ih)})u_{2i}^{\frac{h}{2}} +$$
$$+(\frac{\mu_i^h}{\mu_{2i}^{\frac{h}{2}}}e^{-\frac{\rho}{2}a((2i+1)\frac{h}{2})} + \frac{\mu_i^h}{\mu_{2i+1}^{\frac{h}{2}}})f(ih) + \frac{h}{2}\frac{\mu_i^h}{\mu_{2i+1}^{\frac{h}{2}}}f'(\xi).$$

Using this and (7), expression (6) becomes

$$\mu_i^h(e^{-\frac{\rho}{2}(a((2i+1)\frac{h}{2})+a(ih))} - e^{-\rho a(ih)})u_{2i}^{\frac{h}{2}} +$$
$$+(\frac{\mu_i^h}{\mu_{2i}^{\frac{h}{2}}}e^{-\frac{\rho}{2}a((2i+1)\frac{h}{2})} + \frac{\mu_i^h}{\mu_{2i+1}^{\frac{h}{2}}} - 1)f(ih) + \frac{h}{2}\frac{\mu_i^h}{\mu_{2i+1}^{\frac{h}{2}}}f'(\xi).$$

We now bound separately each of the three terms in this expression. Writing $t_i = ih$, the coefficient of $u_{2i}^{\frac{h}{2}}$ is

$$C_1 = \mu_i^h e^{-\rho a(t_i)}(e^{-\frac{\rho}{2}(a(t_i+\frac{h}{2})-a(t_i))} - 1).$$

Consider the expression in the brackets as a function

$$F(\rho) = e^{-\frac{\rho}{2}(a(t_i+\frac{h}{2})-a(t_i))} - 1$$

and expand it in a two-term Taylor series about the point $\rho = 0$, to obtain

$$F(\rho) = F(0) + \rho\frac{dF}{d\rho}(\xi)$$

for some ξ, $0 < \xi < \rho$. Noting that $F(0) = 0$, this leads to

$$C_1 = \mu_i^h e^{-\rho a(t_i)}(\frac{-\rho}{2})(a(t_i+\frac{h}{2}) - a(t_i))e^{\frac{t_i}{2}(a(t_i+\frac{h}{2})-a(t_i))},$$

Since $a(t_i + \frac{h}{2}) - a(t_i) = \frac{h}{2} a'(t_i + \eta)$, for some $\eta,\ 0 < \eta < \frac{h}{2}$, it follows that

$$|C_1| \leq Ch\mu_i^h e^{-\rho a(t_i)} \rho e^{-\frac{1}{2}\xi a(t_i + \frac{h}{2})} e^{\frac{1}{2}a(t_i)}.$$

Also $0 < \xi < \rho$ and $e^{-\frac{1}{2}\xi a(t_i + \frac{h}{2})} \leq 1$, and so we have

$$|C_1| \leq Ch\mu_i^h e^{-\frac{\rho}{2}a(t_i)} = Ch\frac{\frac{1}{2}\rho a(t_i)}{sinh(\frac{1}{2}\rho a(t_i))} \leq Ch, \tag{10}$$

because $\frac{y}{sinh(y)}$ is bounded for all real y.
Likewise, it is not hard to see that the coefficient of $f(ih)$ can be written in the form

$$C_2 = -C_1 + \frac{\mu_i^h}{a(t_i)} \frac{a(t_i) - a(t_i + \frac{h}{2})}{a(t_i + \frac{h}{2})} (1 - e^{-\frac{\rho}{2}a(t_i + \frac{h}{2})}).$$

The second expression here can be written as

$$\frac{h}{2} \frac{\mu_i^h}{a(t_i)} \frac{a'(t_i + \eta)}{a(t_i + \frac{h}{2})} (1 - e^{-\frac{\rho}{2}a(t_i + \frac{h}{2})}) \leq Ch \frac{1 - e^{-\rho a(t_i + \frac{h}{2})}}{1 - e^{-\rho a(t_i)}} \leq C'h,$$

and it follows that

$$|C_2| \leq |C_1| + C'h \leq Ch \tag{11}$$

as required.
Finally, the coefficient of f' is

$$C_3 = \frac{h}{2} \frac{1 - e^{-\rho a(t_i + \frac{h}{2})}}{1 - e^{-\rho a(t_i)}} \leq Ch. \tag{12}$$

Combining the separate estimates (10), (11) and (12) we conclude that

$$|L_\varepsilon^h (u_{2i}^{\frac{h}{2}} - u_i^h)| \leq Ch. \tag{13}$$

Using Lemma 1.4 we then obtain

$$|u_{2i}^{\frac{h}{2}} - u_i^h| \leq Ch,$$

which is an estimate, at the common mesh points, of the difference between the discrete solutions on the two meshes.

Suppose now that, for any function F, $|F(h) - F(\frac{h}{2})| \leq Ch^p$ for all $h, 0 < h \leq h_0$, and some positive number p. Then, using the hypothesis repeatedly, we get, for all positive integers n,

$$|F(h) - F(\frac{h}{2^n})| = |(F(h) - F(\frac{h}{2})) + (F(\frac{h}{2}) - F(\frac{h}{2^2})) + ... + (F(\frac{h}{2^{n-1}}) - F(\frac{h}{2^n}))|$$
$$\leq Ch^p(1 + \frac{1}{2^p} + ... + \frac{1}{2^{(n-1)p}}) < \frac{Ch^p}{1 - \frac{1}{2^p}}.$$

Taking the limit, as $n \to \infty$, we get

$$|F(h) - F(0)| \leq Ch^p \frac{1}{1 - \frac{1}{2^p}}.$$

Applying this result to $F(h) = u_\varepsilon^h(t_i)$, with $p = 1$, we conclude from (13) and the consistency and stability of (P_ε^h), that

$$|(u_\varepsilon^h - u_\varepsilon)(t_i)| \leq Ch.$$

Therefore (P_ε^h) is a parameter-uniform fitted operator method for solving the singular perturbation problem (P_ε).

3 Fitted Mesh Methods

The aim of this section is to use the fitted mesh method to solve the simple problem (P_ε) introduced above and to prove that it is parameter-uniform. We use a general method of proof, deliberately avoiding any simplifications that could arise due to the simplicity of the problem. Our purpose is to illustrate the general method of proof in this simple setting.

The fitted mesh method (P_ε^h) consists of the backward Euler finite difference operator on a specially constructed non-uniform mesh Ω_ε^N

$$L_\varepsilon^h U_\varepsilon^N(t_i) \equiv \varepsilon D^- U_\varepsilon^N(t_i) + a(t_i) U_\varepsilon^N(t_i) = f(t_i), \quad t_i \in \Omega_\varepsilon^N,$$
$$u_\varepsilon^h(0) = u_\varepsilon(0),$$

where Ω_ε^N is a carefully constructed piecewise uniform mesh with a transition parameter σ due to Shishkin (note that this σ should not be confused with the fitting factor σ in the previous section). This piecewise uniform mesh is constructed by first specifying a σ in the interior of $\overline{\Omega}$ and then constructing two uniform meshes on each of the two resulting subintervals of $\overline{\Omega}$. Taking N to be an even number, each of these uniform meshes has $\frac{N}{2}$ equal mesh intervals and their common point σ is called the transition point between these two uniform meshes. If the point $\sigma = \frac{1}{2}$, it is clear that the complete mesh $\overline{\Omega}^N = \{t_i\}_{i=0}^N$ on T is a uniform mesh, but if $\sigma \neq \frac{1}{2}$, say $\sigma < \frac{1}{2}$, then the mesh in $(0, \sigma]$ is finer than the mesh in $(\sigma, T]$. The explicit definition of the Shishkin mesh Ω_ε^N in this case is as follows.
First, the transition point is defined by

$$\sigma = \min(\frac{1}{2}, \varepsilon \ln N) \tag{14}$$

and it is not hard to see that the mesh widths of the fine and coarse uniform meshes are

$$h = \frac{2\sigma}{N} \quad \text{and} \quad H = \frac{2(1-\sigma)}{N}, \tag{15}$$

respectively. Then, the mesh points are given explicitly by

$$t_i = ih, \quad i = 0 : \frac{N}{2}, \tag{16}$$

$$t_i = \sigma + (i - \frac{N}{2})H, \quad i = \frac{N}{2} + 1 : N. \tag{17}$$

Notice that the Shishkin mesh Ω_ε^N depends on N and ε.
The main theoretical result we want to prove is contained in the following theorem, which provides a parameter-uniform error estimate of essentially first order, in the sense that the $\ln N$ factor means that it is not strictly first order. However, in practice, the factor $\ln N$ is negligible.

Theorem 1.1 *The numerical solutions U_ε^N of (P_ε^N) and the exact solution u_ε of (P_ε) satisfy the following ε-uniform error estimate, for all $N \geq 4$,*

$$\sup_{0<\varepsilon\leq 1} |U_\varepsilon^N - u_\varepsilon|_{\Omega_\varepsilon^N} \leq CN^{-1} \ln N, \tag{18}$$

where C is a constant independent of ε and N.

This will be proved in what follows by a sequence of lemmas.
First we define the reduced problem (P_0) corresponding to (P_ε) as

$$a(t)u_0(t) = f(t). \tag{19}$$

This is obtained by putting $\varepsilon = 0$ in (P_ε). Its solution is clearly

$$u_0(t) = \frac{f(t)}{a(t)}.$$

We then define the Shishkin decomposition of u_ε. We write

$$u_\varepsilon = v_\varepsilon + w_\varepsilon,$$

where v_ε is the smooth component of the decomposition and is defined to be the solution of the problem

$$L_\varepsilon v_\varepsilon = f, \quad v_\varepsilon(0) = u_0(0);$$

u_0 being the solution of the reduced problem. It follows that the singular component w_ε must be the solution of the problem

$$L_\varepsilon w_\varepsilon = 0, \quad w_\varepsilon(0) = u_\varepsilon(0) - v_\varepsilon(0).$$

This decomposition enables us to obtain sharper bounds than the classical bounds given in Lemma 1.5.

Note that the equation for v_ε gives $\varepsilon v'_\varepsilon(0) + a(0)v_\varepsilon(0) = f(0)$ and the initial condition for v_ε gives $a(0)v_\varepsilon(0) = f(0)$. Combining these we obtain $\varepsilon v'_\varepsilon(0) = 0$ and so $v'_\varepsilon(0) = 0$.

Note also that $L_\varepsilon v'_\varepsilon = \varepsilon v''_\varepsilon + a v'_\varepsilon$ and, by differentiating the equation satisfied by v_ε, we get $\varepsilon v''_\varepsilon + (a v_\varepsilon)' = f'$. Eliminating v''_ε gives $L_\varepsilon v'_\varepsilon = f' - a' v_,$.
The sharper bounds are contained in the following lemma.

Lemma 1.6 *The components v_ε, w_ε of the exact solution u_ε satisfy the bounds*

$$|v_\varepsilon|_k \le C, \quad k = 0, 1 \text{ and } |v_\varepsilon|_2 \le C\varepsilon^{-1},$$
$$|w_\varepsilon^{(k)}(t)| \le C\varepsilon^{-k} e^{-\frac{\alpha t}{\varepsilon}}, \quad k = 0, 1, 2 \text{ for all } t \in \overline{\Omega}.$$

Proof The bound on v_ε follows from the corresponding bound in Lemma 1.5, since v_ε is the solution to a problem of the same form as (P_ε).
The bound on v'_ε is obtained by considering the two functions $\phi^\pm(t) = C(1 + |f'|) \pm v'_\varepsilon(t)$. Then, $\phi^\pm(0) = C(1 + |f'|) \pm v'_\varepsilon(0) \ge v'_\varepsilon(0) = 0$. Furthermore, $L_\varepsilon \phi^\pm(t) = a(t)C(1 + |f'|) \pm L_\varepsilon v'_\varepsilon(t) \ge C\alpha(1 + |f'|) \pm (f' - a' v_,) \ge 0$. The required bound on v'_ε follows from the bound on v_ε and the maximum principle Lemma 1.1. The bound on v''_ε is obtained from the equation $\varepsilon v''_\varepsilon + (a v_\varepsilon)' = f'$ and the bounds on v_ε and v'_ε.

The bound on w_ε is obtained similarly. We introduce the functions

$$\psi^\pm(t) = Ce^{-\frac{\alpha t}{\varepsilon}} \pm w_\varepsilon(t),$$

where C is a suitably large constant. Then

$$\psi^\pm(0) = C \pm w_\varepsilon(0) = C \pm (u_\varepsilon(0) - v_\varepsilon(0)) \ge C - (|u_0| + |v_0(0)|) \ge 0.$$

Also,

$$L_\varepsilon \psi^\pm(t) = CL_\varepsilon e^{-\frac{\alpha t}{\varepsilon}} \pm L_\varepsilon w_\varepsilon(t) = CL_\varepsilon e^{-\frac{\alpha t}{\varepsilon}} = (a(t) - \alpha)e^{-\frac{\alpha t}{\varepsilon}} > 0.$$

From the maximum principle we then have

$$\psi^\pm(t) \ge 0 \text{ for all } t \in \overline{\Omega},$$

and so

$$|w_\varepsilon^{(k)}(t)| \le C\varepsilon^{-k}e^{-\frac{\alpha t}{\varepsilon}} \quad \text{for all} \quad t \in \overline{\Omega},$$

as required.
To bound the derivatives of w_ε we use the differential equation repeatedly. We have

$$w_\varepsilon'(t) = -\varepsilon^{-1}a(t)w_\varepsilon(t)$$

and so

$$|w_\varepsilon'(t)| = |\varepsilon^{-1}a(t)w_\varepsilon(t)| \le C\varepsilon^{-1}e^{-\frac{\alpha t}{\varepsilon}}.$$

Similarly

$$w_\varepsilon''(t) = -\varepsilon^{-1}(a(t)w_\varepsilon'(t) + a'(t)w_\varepsilon(t))$$

and so

$$|w_\varepsilon''(t)| \le C\varepsilon^{-2}e^{-\frac{\alpha t}{\varepsilon}},$$

as required.

This completes the proof of this lemma.

The Shishkin decomposition of the discrete solution is analogous to that of the exact solution. We have

$$U_\varepsilon^N = V_\varepsilon^N + W_\varepsilon^N,$$

where V_ε^N is the smooth component of the decomposition and is defined to be the solution of the problem

$$L_\varepsilon^N V_\varepsilon^N = f, \quad V_\varepsilon^N(0) = v_\varepsilon(0).$$

It follows that the singular component W_ε^N must be the solution of the problem

$$L_\varepsilon W_\varepsilon^N = 0, \quad W_\varepsilon^N(0) = w_\varepsilon(0).$$

The error can now be decomposed into smooth and singular components as follows

$$U_\varepsilon^N - u_\varepsilon = (V_\varepsilon^N - v_\varepsilon) + (W_\varepsilon^N - w_\varepsilon). \tag{20}$$

These are now bounded separately. We bound the smooth component first. We have

$$L_\varepsilon^N(V_\varepsilon^N - v_\varepsilon) = f - L_\varepsilon^N v_\varepsilon = (L_\varepsilon - L_\varepsilon^N)v_\varepsilon = \varepsilon(\frac{d}{dt} - D^-)v_\varepsilon.$$

Hence, using (4), we have, for each $t_i \in \Omega_\varepsilon^N$,

$$|L_\varepsilon^N(V_\varepsilon^N - v_\varepsilon)(t_i)| = |\varepsilon(\frac{d}{dt} - D^-)v_\varepsilon(t_i)| \le \frac{\varepsilon}{2}(t_i - t_{i-1})|v_\varepsilon|_{2,[t_{i-1},t_i]}.$$

Noting that, for a Shishkin mesh, $t_i - t_{i-1} \leq 2N^{-1}$ and using the bound for $|v_\varepsilon|_2$ in the Lemma 1.6, we obtain

$$|L_\varepsilon^N(V_\varepsilon^N - v_\varepsilon)(t_i)| \leq C\varepsilon N^{-1}\varepsilon^{-1} = CN^{-1}.$$

From the stability of L_ε^N it follows that

$$|(V_\varepsilon^N - v_\varepsilon)(t_i)| \leq CN^{-1}, \text{ for all } t_i \in \Omega_\varepsilon^N,$$

which is

$$|V_\varepsilon^N - v_\varepsilon|_{\Omega_\varepsilon^N} \leq CN^{-1}, \tag{21}$$

the required bound on the smooth component of the error.
For the singular component of the error we need to use the two different estimates of the local truncation error. From (4) we obtain

$$|L_\varepsilon^N(W_\varepsilon^N - w_\varepsilon)(t_i)| = |\varepsilon(\frac{d}{dt} - D^-)w_\varepsilon(t_i)| \leq \frac{\varepsilon}{2}(t_i - t_{i-1})|w_\varepsilon|_{2,[t_{i-1},t_i]} \tag{22}$$

and similarly from (3)

$$|L_\varepsilon^N(W_\varepsilon^N - w_\varepsilon)(t_i)| = |\varepsilon(\frac{d}{dt} - D^-)w_\varepsilon(t_i)| \leq 2\varepsilon|w_\varepsilon|_{1,[t_{i-1},t_i]}. \tag{23}$$

We consider the following two possibilities separately; either $\sigma = \frac{1}{2}$ or $\sigma = \frac{\varepsilon}{\alpha}\ln N$. In the first case, since $\sigma = \min\{\frac{1}{2}, \frac{\varepsilon}{\alpha}\ln N\} = \frac{1}{2}$, we have $\frac{1}{2} \leq \frac{\varepsilon}{\alpha}\ln N$ or $\varepsilon^{-1} \leq \frac{2}{\alpha}\ln N$. Since, in this case, the mesh is uniform $t_i - t_{i-1} = N^{-1}$ and using the bound for $|w_\varepsilon|_{2,[t_{i-1},t_i]}$ in Lemma 1.4, we obtain

$$|L_\varepsilon^N(W_\varepsilon^N - w_\varepsilon)(t_i)| \leq C\varepsilon N^{-1}\varepsilon^{-2}e^{-\frac{\alpha t_{i-1}}{\varepsilon}} \leq CN^{-1}\varepsilon^{-1} \leq CN^{-1}\ln N.$$

From the stability of L_ε^N it follows that

$$|(W_\varepsilon^N - w_\varepsilon)(t_i)| \leq CN^{-1}\ln N, \text{ for all } t_i \in \Omega_\varepsilon^N, \tag{24}$$

which is the required bound.

In the second case the argument depends on the location of the mesh point t_i. There are 3 possibilities:

- $t_i \in (0, t_{\frac{N}{2}}]$, in the fine mesh;
- $t_i \in (t_{\frac{N}{2}+1}, 1]$, in the coarse mesh;
- $t_i = t_{\frac{N}{2}} = \sigma$, at the transition point.

and we recall that for all of these we are dealing with the case $\sigma = \frac{\varepsilon}{\alpha}\ln N$. We consider each possibility in turn.

We begin with the fine mesh. Since $t_i - t_{i-1} = \frac{2\sigma}{N} = \frac{2\varepsilon}{\alpha} N^{-1} \ln N$, using the bound for $|w_\varepsilon|_{2,[t_{i-1},t_i]}$ in Lemma 1.4, we obtain from (22)

$$|L_\varepsilon^N (W_\varepsilon^N - w_\varepsilon)(t_i)| \leq C\varepsilon^2 N^{-1} \ln N \, \varepsilon^{-2} e^{-\frac{\alpha t_{i-1}}{\varepsilon}} \leq CN^{-1} \ln N. \tag{25}$$

On the other hand, in the coarse mesh $t_i > t_{i-1} \geq \sigma = \frac{\varepsilon}{\alpha} \ln N$ and so

$$e^{-\frac{\alpha t_{i-1}}{\varepsilon}} \leq e^{-\frac{\alpha\sigma}{\varepsilon}} = e^{-\ln N} = N^{-1}.$$

Then, using the estimate (23) of the local truncation error and the bound for $|w_\varepsilon|_{1,[t_{i-1},t_i]}$ in Lemma 1.4, we obtain

$$|L_\varepsilon^N (W_\varepsilon^N - w_\varepsilon)(t_i)| \leq 2\varepsilon |w_\varepsilon|_{1,[t_{i-1},t_i]} \leq Ce^{-\frac{\alpha t_{i-1}}{\varepsilon}} \leq CN^{-1}. \tag{26}$$

Finally, at the transition point, $t_{i-1} = \sigma - 2\sigma N^{-1}$, we have

$$e^{-\frac{\alpha t_{i-1}}{\varepsilon}} = e^{-\frac{\alpha\sigma}{\varepsilon}} e^{\frac{2\alpha\sigma N^{-1}}{\varepsilon}} = e^{-\ln N} e^{2N^{-1}\ln N} = N^{-1} e^{\ln(N^{2N^{-1}})} = N^{-1} N^{2N^{-1}} = N^{-1} (N^{\frac{1}{N}})^2 \leq CN^{-1},$$

since $N^{\frac{1}{N}}$ is bounded. Then, using the estimate (23) of the local truncation error and the bound for $|w_\varepsilon|_{1,[x_{i-1},x_i]}$ in Lemma 1.4, we obtain

$$|L_\varepsilon^N (W_\varepsilon^N - w_\varepsilon)(\sigma)| \leq C\varepsilon |w_\varepsilon|_{1,[\sigma-2\sigma N^{-1},\sigma]} \leq Ce^{-\frac{\alpha(\sigma-2\sigma N^{-1})}{\varepsilon}} \leq CN^{-1}. \tag{27}$$

Combining (24), (26) and (27) we see that the singular component of the error satisfies

$$|W_\varepsilon^N - w_\varepsilon|_{\Omega_\varepsilon^N} \leq CN^{-1} \ln N. \tag{28}$$

The bound on the error is obtained by combining the bounds (21) and (28) respectively on the smooth and singular components, which completes the proof of the theorem.

4 Computations

For the numerical solution of differential equations the following codes are useful and widely available: MatLab, Octave and Python-Anaconda. The first is a commercial code with a reduced price for academics and a further reduced price for the student edition. The second and third are open source and free.
In this section we consider another simple initial value problem. This involves the following second order differential equation with constant coefficients

$$m\frac{d^2y}{dt^2} + k\frac{dy}{dt} + cy = 0,$$

which describes the damped vibrations of a point mass m on a spring with spring constant c and damping coefficient k. Here, m, k, c are taken to be positive constants. The initial conditions are

$$y(0) = y_0,$$
$$\frac{dy(0)}{dt} = y_1,$$

where the initial displacement y_0 and the initial velocity y_1 are given.
The exact solution is

$$y(t) = Ae^{\lambda^+ t} + Be^{\lambda^- t},$$

where $A,\ B$ are integration constants and $\lambda^+,\ \lambda^-$ are the roots of

$$m\lambda^2 + k\lambda + c = 0,\quad \lambda^{\pm} = \frac{-k \pm \sqrt{k^2 - 4mc}}{2m}.$$

We assume, henceforth, that

$$k^2 > 4mc,$$

which ensures that both of these roots are real and distinct.
The integration constants A, B are determined by the initial conditions as follows:

$$y(0) = 0 \text{ implies that } A + B = 0 \text{ or } B = -A$$

and

$$\frac{dy(0)}{dt} = y_1 \text{ implies that } A(\lambda^+ - \lambda^-) = y_1 \text{ or } A = \frac{y_1}{\lambda^+ - \lambda^-}.$$

This example becomes a singularly perturbed problem if m is small. We write

$$m = \varepsilon, \qquad 0 < \varepsilon << 1.$$

where ε is a singular perturbation parameter multiplying the highest derivative term. The equation is then

$$\varepsilon\frac{d^2y}{dt^2} + k\frac{dy}{dt} + cy = 0. \tag{29}$$

The appropriate initial conditions are now

$$y(0) = 0, \tag{30}$$
$$\varepsilon\frac{dy(0)}{dt} = \gamma, \tag{31}$$

where γ is independent of ε. These differ from the unperturbed case, because the derivative in the second condition is now multiplied by ε. This is necessary here because the first derivative of the solution $|\frac{dy}{dt}|$ is unbounded as $\varepsilon \to 0$, while $|\varepsilon\frac{dy}{dt}|$ is bounded. This is seen at once from the exact solution

$$y(t) = \frac{\gamma}{\varepsilon(\lambda^+ - \lambda^-)}(e^{\lambda^+ t} - e^{\lambda^- t}),$$

where $\lambda^{\pm}$ are the roots of

$$\varepsilon\lambda^2 + k\lambda + c.$$

We assume that $k^2 > 4\varepsilon c$ to ensure that the roots are real and distinct.
The plot of the exact solution for the moderate value $\varepsilon = \frac{1}{2}$ is given in Fig. 1.
Let us investigate now what happens as $\varepsilon \to 0$. The plots of the exact solution for $\varepsilon = \frac{1}{4}, \frac{1}{8}, \frac{1}{16}, \frac{1}{32}$ are given in Fig. 2. As expected we see that as $\varepsilon \to 0$ the gradient of the solution becomes increasingly steep in a neighbourhood of $t = 0$, so we have a layer there. Even for the moderately small value $\frac{1}{32}$, the exact solution near the origin is graphically indistinguishable from the vertical axis.

We now use numerical methods to solve problems with layers. We shall discover that standard finite difference methods are not reliable for this task. The same can be shown to be true for standard finite element methods. By this we mean that they must be used with caution and particular attention must be paid to the relative sizes of the singular perturbation parameter and the mesh parameters.
The finite difference method with centered operator for the above problem is

$$\varepsilon\delta^2 U_i + kD^0 U_i + cU_i = 0, \quad U_0 = 0, \quad \varepsilon D^+ U_0 = \gamma,$$

where the mesh is uniform with mesh spacing $h = \frac{1}{N}$ and N is the number of mesh subintervals in a unit interval on the t-axis.
We look for a numerical solution for $t \in [0, T]$, where, to be specific, we take $T = 3$. Plots of the numerical solution for $\varepsilon = \frac{1}{8}$, $N = 16$ and the corresponding exact solution are given in Fig. 3. The exact solution is plotted as a continuous line, while the values of the numerical solution are denoted by a small circle at each mesh point. It is easy to see that the numerical solution is highly inaccurate.

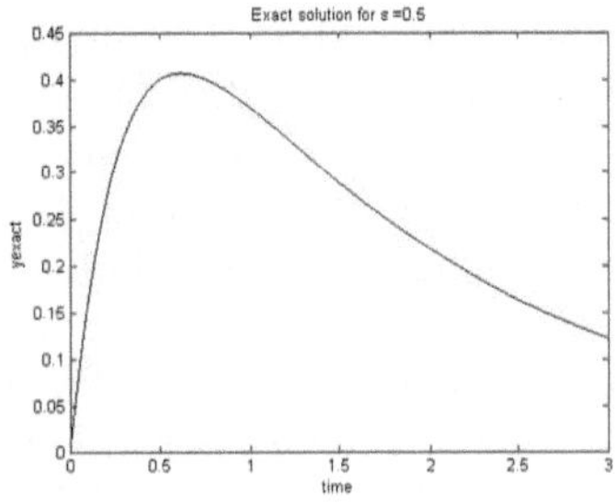

Fig. 1 The exact solution for $\varepsilon = \frac{1}{2}$

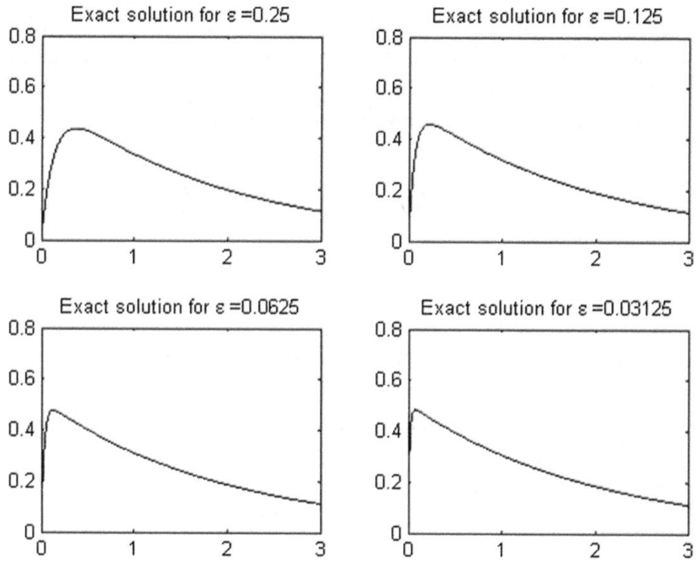

Fig. 2 The exact solution for decreasing values of ε

Fig. 3 The numerical solution with centered operator, uniform mesh for $\varepsilon = \frac{1}{8}$, $N = 16$

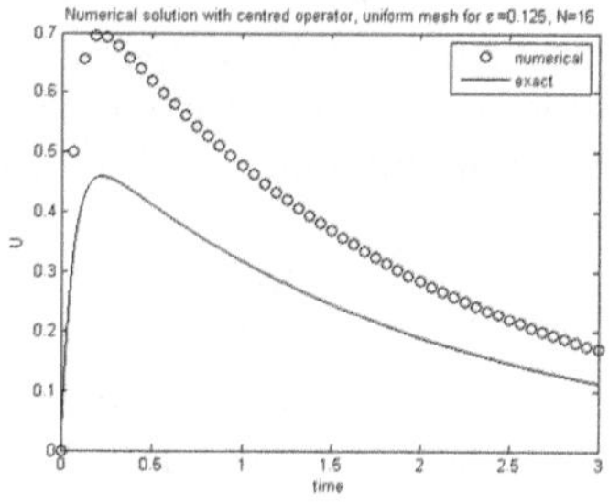

Can this situation be improved and, if so, how? What about a finer mesh? Can you predict what will happen? We take N $= 256$ and plot the results in Fig. 4. These results are a big improvement. But are they robust in the sense that they are just as good for other values of the parameter? So far, ε has had moderate values. Let us try taking the smaller value $\varepsilon = \frac{1}{1000}$. The results are plotted in Fig. 5. We see that the numerical solution is again highly inaccurate outside the boundary layer. Moreover, in a neighbourhood of the boundary layer at $t = 0$ it displays enormous spurious oscillations unrelated to the exact solution. In fact here, the numerical solution is completely different from the exact solution. This is seen more clearly when we restrict the plot of the solution to a neighbourhood of the boundary layer as in Fig. 6.

Can we do anything to overcome these problems? The obvious thing to try is to refine the mesh as before. We therefore take $N = 1024$ and plot the results in Fig. 7. We see there that the spurious oscillations in the neighbourhood of the boundary layer have been eliminated, but there remain large errors in the rest of the domain.

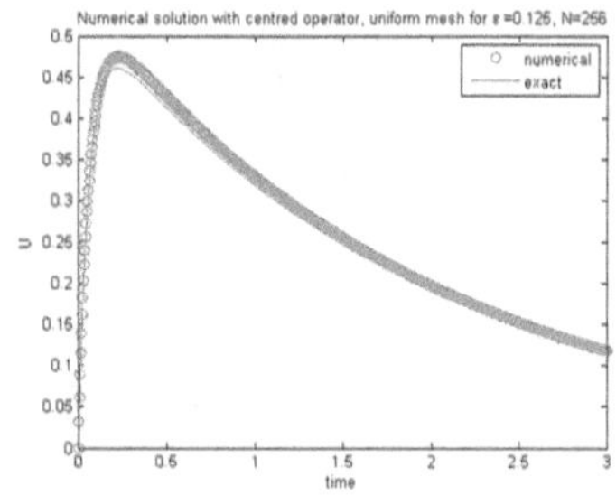

Fig. 4 The numerical solution with centered operator, uniform mesh for $\varepsilon = \frac{1}{8}, \quad N = 256$

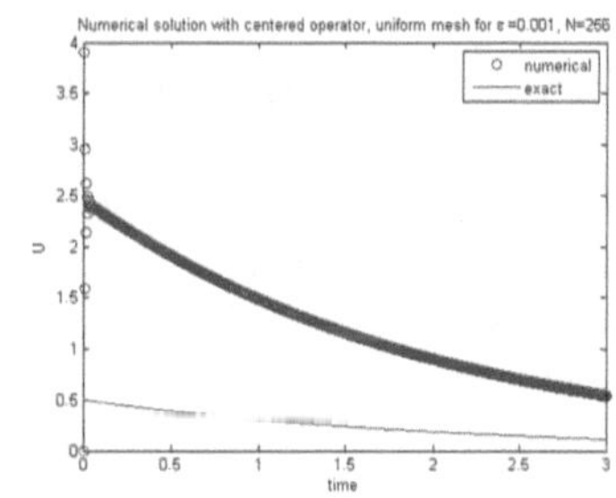

Fig. 5 The numerical solution with centered operator, uniform mesh for $\varepsilon = \frac{1}{1000}, \quad N = 256$

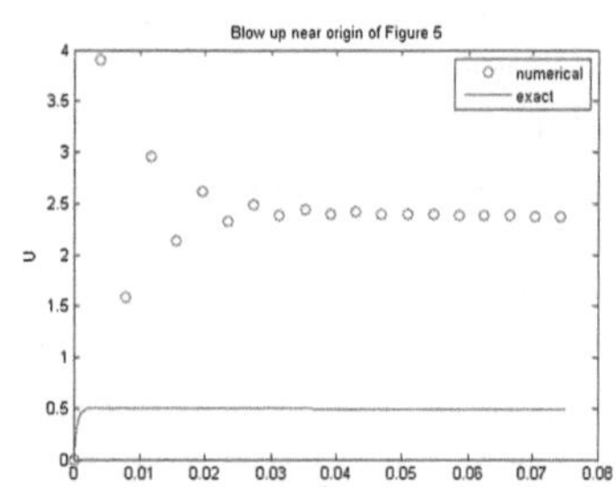

Fig. 6 Blow up for centered operator, uniform mesh for $\varepsilon = \frac{1}{1000}, \quad N = 256$

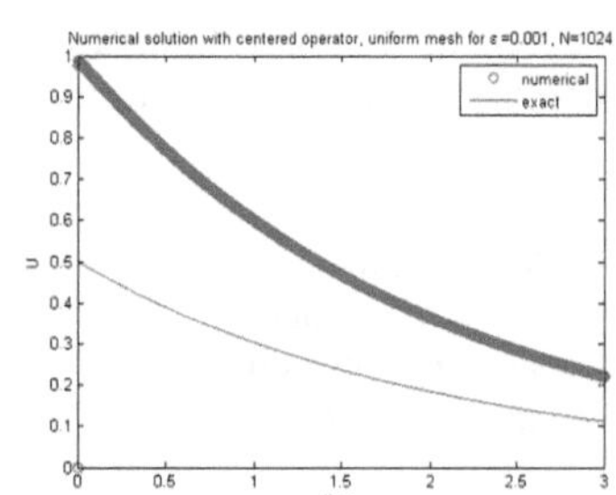

Fig. 7 The numerical solution with centered operator, uniform mesh for $\varepsilon = \frac{1}{1000}, \quad N = 1024$

Our overall conclusion is that the finite difference method with centered operator on a uniform mesh is unreliable for this problem, even for moderate values of ε. We can also say that this example has illustrated the fact that numerical experimentation is an extremely useful tool for distinguishing between good and bad numerical methods.

We now try a different numerical method, namely, the following *upwind* method

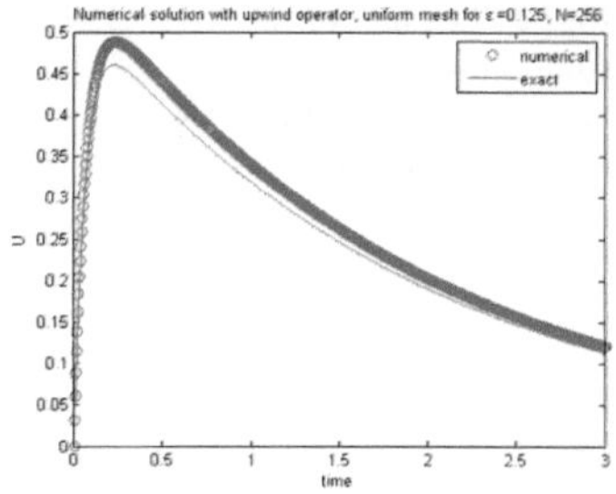

Fig. 8 The numerical solution with upwind operator, uniform mesh for $\varepsilon = \frac{1}{8}$, $N = 256$

$$\varepsilon\delta^2 U_i + kD^+ U_i + cU_i = 0, \quad U_0 = 0, \quad \varepsilon D^+ U_0 = \gamma.$$

Here the first order derivative is replaced by a one-sided finite difference quotient and the method is said to be upwinded. The mesh is uniform with N equally spaced mesh subintervals. The results of applying this method for $\varepsilon = \frac{1}{8}$ and $N = 256$ are shown in Fig. 8. The solution is satisfactory, although not quite as accurate as the solution given by the previous method with the centered difference operator.

Let us now look at the behaviour for smaller $\varepsilon = \frac{1}{1000}$, keeping the same uniform mesh with N = 256. The results are shown in Fig. 9. Again the errors have become enormous, when ε is reduced. The one positive thing about this method is that the numerical solution shows no sign of spurious oscillations, as was the case for the method with the centered operator. Our overall conclusion is that the finite difference method with an upwind operator on a uniform mesh is again unreliable for this problem.

It has become apparent that we need a method that produces numerical solutions with an accuracy that is not destroyed when ε becomes small. To achieve this we use a parameter-uniform numerical method. For the singularly perturbed initial value problem (29), (30), (31) it can be shown theoretically that a parameter-uniform method is obtained by using the upwind finite difference operator described above together with the Shishkin mesh that was constructed in the previous section.

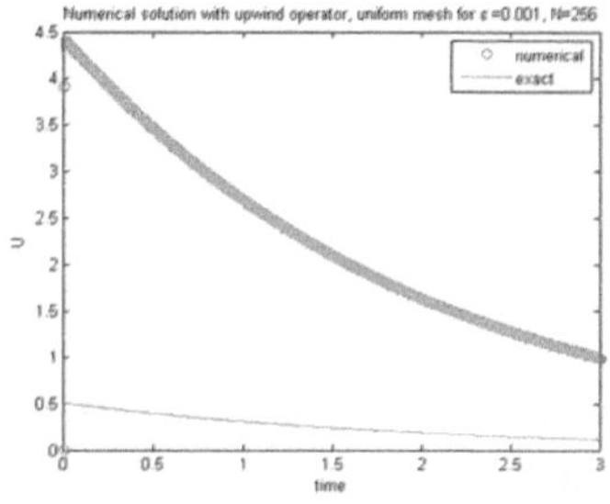

Fig. 9 The numerical solution with upwind operator, uniform mesh for $\varepsilon = \frac{1}{1000}$, $N = 256$

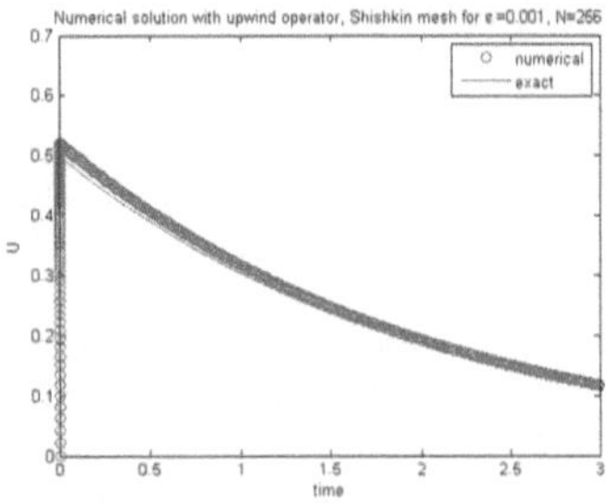

Fig. 10 The numerical solution with upwind operator, Shishkin mesh for $\varepsilon = \frac{1}{1000}$, $N = 256$

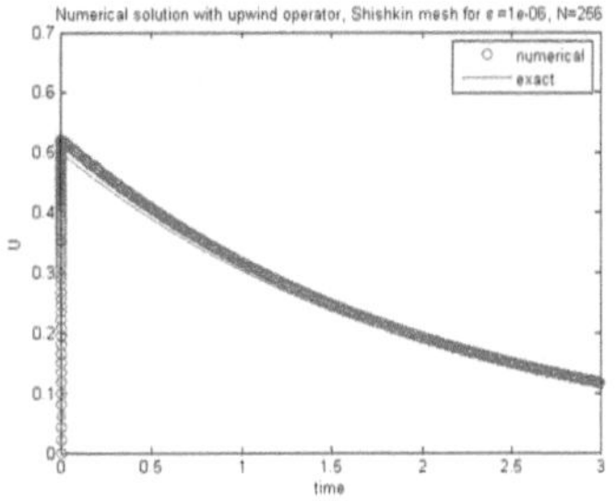

Fig. 11 The numerical solution with upwind operator, Shishkin mesh for $\varepsilon = 10^{-6}$, $N = 256$

The numerical results obtained by applying this parameter-uniform method are plotted in Fig. 10. We see that the agreement with the exact solution is excellent for this small value of ε. The use of the Shishkin mesh has eliminated the large errors of the methods based on the centered or upwind operators on uniform meshes.

It remains to see whether further reduction of ε destroys the numerical solution or not. We take $\varepsilon = 10^{-6}$, much smaller than any previous choice. We see in Fig. 11 that the numerical results with this method are brilliant; just as good as for $\varepsilon = \frac{1}{1000}$ and without the need for any finer mesh than was used in the previous case!

In this section we have shown experimentally that standard finite difference methods, applied on uniform meshes, are not reliable for solving singularly perturbed differential equations, when the singular perturbation parameter becomes small. To overcome this problem it may be possible to increase the number of mesh subintervals so that N is of order $\frac{1}{\varepsilon}$. This brute force approach can be used for simple problems, but for very large systems of equations and very small values of the singular perturbation parameters this may not be feasible. In such cases the use of suitable generalisations (for example in the books [1–5]) of the parameter-uniform methods described in this tutorial may offer an elegant and reliable option.

References

1. E.P. Doolan, J.J.H. Miller, W.H.A. Schilders, *Uniform Numerical Methods for Problems with Initial and Boundary Layers* (Boole Press, Dublin, 1980)
2. P.A. Farrell, A. Hegarty, J.J.H. Miller, E. O'Riordan, G.I. Shishkin, *Robust Computational Techniques for Boundary Layers*, Chapman & Hall/CRC Press (2000)
3. J.J.H. Miller, E. O'Riordan, G.I. Shishkin, *Fitted Numerical Methods for Singular Perturbation Problems* (World Scientific Publishing Co., Singapore, New Jersey, London, Hong Kong, 2012) Revised edition
4. H.-G. Roos, M. Stynes, L. Tobiska, *Robust Numerical methods for Singularly Perturbed Differential Equations*, 2nd edn. Springer (2008)
5. G.I. Shishkin, L.P. Shishkina, *Difference Methods for Singular Perturbation Problems*, Chapman & Hall/CRC Press (2009)

Interior Layers in Singularly Perturbed Problems

Eugene O'Riordan

Abstract To construct layer adapted meshes for a class of singularly perturbed problems, whose solutions contain boundary layers, it is necessary to identify both the location and the width of any boundary layers present in the solution. Additional interior layers can appear when the data for the problem is not sufficiently smooth.In the context of singularly perturbed partial differential equations, the presence of any interior layer typically requires the introduction of a transformation of the problem, which facilitates the necessary alignment of the mesh to the trajectory of the interior layer. Here we review a selection of published results on such problems to illustrate the variety of ways that interior layers can appear.

Keywords Singular perturbation problems · Finite difference scheme · Shishkin mesh · Interior layers

1 Introduction

The analytical solutions of linear singularly perturbed differential equations typically contain boundary layers. To construct parameter-uniform numerical methods [1] for such problems, the location, width and strength of all layers present in the solution needs to be identified. In addition to boundary layers, interior layers can also appear in certain types of singularly perturbed problems. Interior layers can form for several reasons. For example, interior layers can appear due to the presence of turning points, non-smooth coefficients, non-smooth boundary/initial data, non-linearities or lack of compatibility at any corner points within the domain. In the case of linear problems, the strength and width of any interior layer will depend on whether the problem is of convection-diffusion or reaction-diffusion type on either side of the layer. If turning points are present (when the convective coefficient is continuous and passes through zero at some point within the domain) then the nature of any boundary or interior

E. O'Riordan (✉)
Dublin City University, Dublin 9, Ireland
e-mail: eugene.oriordan@dcu.ie

V. Sigamani et al. (eds.), *Differential Equations and Numerical Analysis*,
Springer Proceedings in Mathematics & Statistics 172,
DOI 10.1007/978-81-322-3598-9_2

layer will depend on the rate at which the convective coefficient approaches its root. The level of smoothness of the problem data will also influence the creation of interior layers. In the case of singularly perturbed parabolic problems, the location of any interior layer may move with time and any layer-adaptive mesh will be required to track this movement. In the case of nonlinear problems, the location of any interior layer may not be known explicitly and detailed asymptotic information about the location of the interior layer will be required in order to construct a suitable mesh for the problem. In this paper, we review some recent results on singularly perturbed problems with interior layers.

In Sect. 2, we begin with some standard results on linear singularly perturbed problems with smooth data, where we identify sufficient regularity for the data so that no internal layer appears in the solution and the standard results on parameter-uniform numerical methods for both convection-diffusion and reaction-diffusion problems immediately apply. In Sect. 3, we introduce a discontinuity into a simple initial value problem and give the modifications in both the mesh and the analysis that are required to retain the basic result of first order parameter-uniform convergence. In Sect. 4, we discuss some results concerning interior layers appearing in linear ordinary differential equations due to discontinuities in the data. In Sect. 5 we briefly consider a particular type of singularly perturbed turning point problem with interior layers of exponential-type in the solution. In Sect. 6, we outline the issues around capturing a moving interior layer and in Sect. 7 we conclude the paper with some comments on interior layers occurring in nonlinear singularly perturbed problems.

Notation. Here and throughout the paper, C is a generic constant independent of both the singular perturbation parameter ε and N, which is the number of mesh elements used in any co-ordinate direction. Also, $\|\cdot\|_D$ denotes the maximum pointwise norm over the set D.

2 Linear Singularly Perturbed Problems with Smooth Data

Linear second order singularly perturbed boundary value problems can be categorized into two broad problem classes: problems of reaction-diffusion or convection-diffusion type.

Consider the following class of singularly perturbed reaction-diffusion problems of the form: Find $u \in C^{k+2}(\bar{\Omega}),\ k \geq 0$ such that

$$L_{\varepsilon,(1)}u := -\varepsilon u'' + b(x)u = f(x), \quad x \in \Omega := (0,1), \tag{1a}$$

$$u(0) = u_0, u(1) = u_1; \quad b, f \in C^k(\Omega); \quad b(x) \geq \beta > 0,\ x \in \overline{\Omega}. \tag{1b}$$

Due to the assumption $b(x) > 0$, the differential operator is inverse-monotone. That is: *For any function* $y \in C^0(\bar{\Omega}) \cap C^2(\Omega)$ *if* $y(0) \geq 0, y(1) \geq 0$ *and* $L_{\varepsilon,(1)}y(x) \geq 0,\ x \in \Omega$ *then* $y(x) \geq 0, \forall x \in \bar{\Omega}$. This property of the differential operator is used extensively to obtain suitable bounds on various components of the solution and in estimating the level of accuracy of any proposed approximation to the solution.

The associated reduced problem is simply $b(x)v_0(x) = f(x)$. Note that the reduced solution v_0 of any singularly perturbed problem is defined in such a way that

$$\|u - v_0\|_{\Omega_0} \le C\varepsilon^p, \quad \text{meas}(\Omega \setminus \Omega_0) = O(\varepsilon^p \ln(1/\varepsilon)), \quad p > 0$$

and that v_0 solves the associated differential equation with ε formally set to zero. In general, the reduced solution will not satisfy all the boundary conditions and, then, boundary layers form when there is a discrepancy between the boundary values of the reduced solution and the solution. To correct this discrepancy, we define the leading term of the two boundary layer functions for problem (1) to be

$$w_{L,0}(x) := (u(0) - v_0(0))e^{-\sqrt{\frac{b(0)}{\varepsilon}}x}, \quad w_{R,0}(x) := (u(1) - v_0(1))e^{-\sqrt{\frac{b(1)}{\varepsilon}}(1-x)}.$$

Then we observe that $L_{\varepsilon,(1)}(v_0 + w_{L,0} + w_{R,0}) = O(\sqrt{\varepsilon})$ as

$$\|L_{\varepsilon,(1)}v_0\|_{\bar{\Omega}} \le C\varepsilon \quad \text{and} \quad |L_{\varepsilon,(1)}w_{L,0}(x)| \le Cxe^{-\sqrt{\frac{b(0)}{\varepsilon}}x} \le C\sqrt{\varepsilon};$$

and $(v_0 + w_{L,0} + w_{R,0})(0) = Ce^{-\sqrt{\frac{b(1)}{\varepsilon}}}$, $(v_0 + w_{L,0} + w_{R,0})(1) = Ce^{-\sqrt{\frac{b(0)}{\varepsilon}}}$. Hence, we have constructed the following asymptotic expansion

$$u_{asy} = v_0 + w_{L,0} + w_{R,0}, \quad \text{where} \quad \|u - u_{asy}\|_{\bar{\Omega}} \le C\sqrt{\varepsilon}.$$

From this bound, we see that $f(x)/b(x)$ is indeed the reduced solution for the reaction-diffusion problem.

If we impose the constraint $\varepsilon \le CN^{-2p}$, $p > 0$, then without performing any numerical calculations the above asymptotic expansion of $v_0 + w_{L,0} + w_{R,0}$, yields an $O(N^{-p})$-approximate solution to the solution u with relatively low regularity assumed (i.e., let $k = 2$ in (1)). However, we wish to generate approximations without imposing a constraint of the form $\varepsilon \le CN^{-2p}$ on the set of problems being considered. Our interest lies in designing parameter-uniform numerical methods [1] for a large set of singularly perturbed problems. Parameter-uniform numerical methods guarantee convergence of the numerical approximations, without imposing a mesh-dependent restriction on the permissible size of the singular perturbation parameter. To establish parameter-uniform asymptotic error bounds on any numerical approximations generated, we will require bounds on the derivatives of the components. The above asymptotic expansion does not yield information about the derivatives of the solution.

In place of the asymptotic expansion, we will utilize a Shishkin decomposition [2] of the solution in the analysis of appropriate numerical methods for (1). Consider the extended domain $\Omega^* := (-L, L)$, $1 < L$ and the extended functions $b^*, f^* \in C^k(\Omega^*)$ of b, f. The extended regular component: $v^* := v_0^* + \varepsilon v_1^*$, where the correction v_1^* to the reduced solution v_0^* satisfies

$$L^*_{\varepsilon,(1)} v^*_1 = (v^*_0)'', \quad v^*_1(-L) = v^*_1(L) = 0.$$

Then the regular component $v \in C^k(\bar{\Omega})$ satisfies the boundary value problem

$$L_{\varepsilon,(1)} v = f, \quad v(0) = v^*(0), \; v(1) = v^*(1).$$

If $v(0) \neq u(0)$, then a boundary layer will be present in a neighbourhood of $x = 0$. Layer components $w_L, w_R \in C^{k+2}(\bar{\Omega})$ at either end point, are defined as the solutions of the following homogeneous problems

$$L_{\varepsilon,(1)} w_L = 0, \; w_L(0) = u(0) - v(0), \; w_L(1) = 0;$$
$$L_{\varepsilon,(1)} w_R = 0, \; w_R(0) = 0, \; w_R(1) = u(1) - v(1).$$

Note that $w_L(x) \neq w_{L,0}(x)$. The Shishkin decomposition (with no remainder term) is now of the form

$$u = v + w_L + w_R.$$

If $b, f \in C^4(\Omega)$ then one can establish [3, Chap. 6] the following bounds on the derivatives of the components of the solution of (1).

$$\|v^{(i)}\|_{\bar{\Omega}} \leq C(1 + \varepsilon^{1-i/2}), \quad i \leq 4;$$
$$|w_L^{(i)}(x)| \leq C\varepsilon^{-i/2} e^{-\sqrt{\frac{\beta}{\varepsilon}}x}, \quad |w_R^{(i)}(x)| \leq C\varepsilon^{-i/2} e^{-\sqrt{\frac{\beta}{\varepsilon}}(1-x)}, \quad i \leq 4.$$

where $v^{(i)}$ denotes the ith-derivative of v. Thus the solution of (1) has boundary layers of width $O(\sqrt{\varepsilon}\log\varepsilon)$ near the end-points $x = 0$ and $x = 1$. Once the location and width of all the layers have been identified (as above) then a layer-adapted mesh can be constructed for the problem.

In the case of convection-diffusion problems of the form: Find $u \in C^{k+2}(\bar{\Omega})$, $k \geq 0$ such that

$$L_{\varepsilon,(2)} u := -\varepsilon u'' + a(x)u' + b(x)u = f(x), \; x \in \Omega, \tag{2a}$$
$$u(0) = u_0, u(1) = u_1; \quad a, b, f \in C^k(\Omega), \quad a(x) \geq \alpha > 0, \; b(x) \geq 0. \tag{2b}$$

The reduced solution satisfies the first order problem: Find $v_0 \in C^{k+1}(\bar{\Omega})$ such that

$$L_{0,(2)} v_0 := a(x)v_0' + b(x)v_0 = f(x), \; x \in \Omega, \quad v_0(0) = u(0).$$

Define the extended domain $\Omega^* := (0, L), \; L > 1$ and $a^*, b^*, f^* \in C^k(\Omega^*)$. The regular component is $v^* := v^*_0 + \varepsilon v^*_1 + \varepsilon^2 v^*_2$, where

$$L^*_{0,(2)} v^*_0 = f^*, \quad v^*_0(0) = u(0); \quad L^*_{0,(2)} v^*_1 = (v^*_0)'', \quad v^*_1(0) = 0;$$
$$L^*_{\varepsilon,(2)} v^*_2 = (v^*_1)'', \quad v^*_2(0) = v^*_2(L) = 0.$$

Then $L^*_{\varepsilon,(2)}v^* = f^*$ and $L_{\varepsilon,(2)}v = f, v(0) = u(0), \quad v(1) = v^*(1)$. If $a, b, f \in C^k(\Omega)$, then $v \in C^k(\bar{\Omega})$. The boundary layer component satisfies the homogeneous problem

$$L_{\varepsilon,(2)}w = 0, \quad w(0) = 0, \ w(1) = u(1) - v(1).$$

Hence $u = v + w$ and if $a, b, f \in C^3(\Omega)$ then [1, Chap. 3]

$$\|v^{(i)}\|_{\bar{\Omega}} \le C(1 + \varepsilon^{2-i}); \quad |w^{(i)}(x)| \le C\varepsilon^{-i}e^{-\alpha(1-x)/\varepsilon}, \quad i \le 3.$$

Using simple stable finite difference scheme with a standard piecewise-uniform Shishkin mesh to produce a numerical approximation U^N, one has for the convection-diffusion problem (2) (see, for example, [1, Chap. 3]): If $a, b, f \in C^3(\Omega)$ then

$$\|u - \bar{U}^N\|_{\Omega} \le CN^{-1}(\ln N)$$

and for the reaction-diffusion problem (1) (see, for example, [3, Chap. 6]): If $b, f \in C^4(\Omega)$ then

$$\|u - \bar{U}^N\|_{\Omega} \le CN^{-2}(\ln N)^2,$$

where $\bar{U}^N$ is the piecewise linear interpolant of the mesh function U^N. Note that, as one would expect, higher regularity is required of the data to establish higher order convergence. The regularity required of the data is dictated by the chosen construction of the solution decomposition. An alternative construction of the solution decomposition can allow one relax the constraints (imposed above) on the data [4].

If there is no modification to the standard numerical method on a layer-adapted mesh, then the above stated orders of parameter-uniform convergence can reduce for less smooth data [2, Sect. 14.2], [5]. If the data is discontinuous, then the standard numerical method will fail to be parameter-uniformly convergent [6, Table 4].

3 Singularly Perturbed Initial Value Problems with Discontinuous Data

Notation: Throughout the paper we adopt the following notation for the jump in a function at an internal point:

$$[f(d)] := f(d^+) - f(d^-), \ 0 < d < 1;$$

and we define the punctured domain by $\Omega_d := (0, 1) \setminus \{d\}$.

To illustrate the effect of discontinuous data, we begin our discussion with a simple initial value problem of the form: Find $u \in C^0(\bar{\Omega}) \cap C^{k+1}(\Omega_d)$ such that

$$\varepsilon u' + a(x)u = f(x), \ a(x) \ge \alpha > 0, \quad x \in \Omega_d, \tag{3a}$$

$$u(0) = u_0, \quad a, f \in C^k(\Omega_d), \quad k \ge 0. \tag{3b}$$

Observe that the differential equation is not applied at the single interior point $x = d$, Instead, the value of the solution at $x = d$ is determined by requiring that the solution be continuous at this internal point. In addition, since $a(x) \geq \alpha > 0,\ x \in \Omega_d$, we have the following useful monotonicity property of the first order differential operator associated with problem (3):

For any function $g \in C^1(\Omega_d) \cap C^0(\bar{\Omega})$, then if $g(0) \geq 0$ and $(\varepsilon g' + ag)(x) \geq 0$ for all $x \in \Omega_d$ then $g(x) \geq 0,\ x \in \bar{\Omega}$.

The solution can be decomposed into a sum $u = v + w + z$, composed of a discontinuous regular component v, a continuous initial layer component w and a discontinuous interior layer component z. The components are defined as the solutions of the problems:

$$\begin{aligned}
\varepsilon v' + a(x)v &= f(x),\ x \in \Omega_d, \quad a(0)v(0) = f(0), \quad a(d^+)v(d^+) = f(d^+);\\
\varepsilon w' + a(x)w &= 0,\ x \in \Omega_d, \quad w(0) = u(0) - v(0), \quad w(d^+) = w(d^-);\\
\varepsilon z' + a(x)z &= 0,\ x \in \Omega_d, \quad z(0) = 0, \quad [z(d)] = -[v(d)].
\end{aligned}$$

Using the above monotonicity property of the differential operator, one can easily deduce the following bounds on these components: If $a, f \in C^1(\Omega_d)$ then

$$\|v^{(i)}\|_{\Omega_d} \leq C(1 + \varepsilon^{1-i}), \qquad i = 0, 1, 2; \tag{4a}$$

$$|w(x)| \leq Ce^{-\alpha x/\varepsilon}, \quad |w^{(i)}(x)|_{\Omega_d} \leq C\varepsilon^{-i}e^{-\alpha x/\varepsilon}, \qquad i = 1, 2; \tag{4b}$$

$$z(x) \equiv 0, \quad x < d, \quad |z(x)| \leq Ce^{-\alpha(d-x)/\varepsilon}, \quad x \in (d, 1); \tag{4c}$$

$$|z^{(i)}(x)|_{(d,1)} \leq C\varepsilon^{-i}e^{-\alpha(d-x)/\varepsilon}, \qquad i = 1, 2. \tag{4d}$$

From these bounds we see that the solution has an initial boundary layer w in the vicinity of $x = 0$ and an interior layer to the right of $x = d$.

Given this *a priori* information, an appropriate distribution of the mesh points $\{x_i\}_{i=0}^N$ is as follows: The end-points of the domain are included as $x_0 = 0,\ x_N = 1$ and the internal point d, where the data is discontinuous, is taken to be the mesh point $x_{N/2}$. The remaining internal mesh points $\omega^N = \{x_i\}_{i=1}^{N/2-1} \cup \{x_i\}_{i=N/2+1}^{N-1}$ are distributed so as to capture the two scales present in the solution. The domain is split into four sub-intervals $[0, 1] = [0, \sigma_1] \cup [\sigma_1, d] \cup [d, d + \sigma_2] \cup [d + \sigma_2, 1]$, where the Shishkin transition parameters [3] are taken to be

$$\sigma_1 := \min\left\{\frac{d}{2}, \frac{\varepsilon}{\alpha}\ln N\right\}, \quad \sigma_2 := \min\left\{\frac{1-d}{2}, \frac{\varepsilon}{\alpha}\ln N\right\}.$$

The mesh elements are distributed equally across these sub-intervals. An appropriate numerical method for this problem is: Find a mesh function U such that:

$$\varepsilon D^- U^N(x_i) + a(x_i^-)U^N(x_i) = f(x_i^-), \quad x_i \in \omega^N \cup \{d\}, \quad U^N(0) = u(0),$$

where D^- is the standard backward finite difference operator. The discrete solution U may be decomposed into the sum $U = V + W + Z$, where the boundary layer component w is approximated by the solution W of the homogeneous problem

$$\varepsilon D^- W(x_i) + a(x_i^-)W(x_i) = 0, \quad x_i \in \omega^N \cup \{d\}, \quad W(0) = u(0) - v(0).$$

The discrete regular component V and discrete interior layer component Z are multi-valued (at $x = d$) functions and are defined to be

$$V := \begin{cases} V^-, & x_i \in [0, d] \\ V^+, & x_i \in [d, 1] \end{cases}, \quad Z := \begin{cases} 0, & x_i \in [0, d] \\ Z^+, & x_i \in [d, 1] \end{cases},$$

where

$$\begin{aligned} \varepsilon D^- V^-(x_i) + a(x_i^-)V^-(x_i) &= f(x_i^-),\ x_i \in (0, d], & V^-(0) &= v(0); \\ \varepsilon D^- V^+(x_i) + a(x_i)V^+(x_i) &= f(x_i),\ x_i \in (d, 1], & V^+(d) &= v(d^+); \\ \varepsilon D^- Z^+(x_i) + a(x_i)Z^+(x_i) &= 0,\ x_i \in (d, 1], & Z^+(d) &= -[V](d). \end{aligned}$$

Based on classical stability, truncation error analysis and the parameter-explicit bounds on the derivatives of w given in (4b), we conclude that in the initial layer region $[0, \sigma_1]$

$$|W(x_i) - w(x_i)|_{[0,\sigma_1]} \le CN^{-1} \ln N.$$

Then we note that, for $x_i \in [\sigma_1, 1]$, the error in the layer component is

$$\begin{aligned} |W(x_i) - w(x_i)|_{[\sigma_1,1]} &\le |W(x_i)| + |w(x_i)| \\ &\le |W_\varepsilon(\sigma_1)| + CN^{-1} \le CN^{-1}, \quad \text{if} \quad 2\sigma_1 < d. \end{aligned}$$

Using the bounds (4a) on the regular component we also conclude that

$$|V(x_i) - v(x_i)| \le CN^{-1}, \quad x_i \in \Omega_d, \quad |V^-(d) - v(d^-)| \le CN^{-1}.$$

Note that $Z^+(d) = -[v(d)] + V^-(d) - v(d^-)$ and by examining the error $|Z - z|$ on $(d, d + \sigma_2]$ and $(d + \sigma_2, 1]$ separately we conclude that

$$|Z(x_i) - z(x_i)|_{[d,1]} \le CN^{-1} \ln N, \quad \text{if} \quad 2\sigma_2 < 1 - d.$$

If $2\sigma_1 = d$ or $2\sigma_2 = 1 - d$ then a standard stability and consistency argument yields

$$\max_i |U^N(x_i) - u(x_i)| \le CN^{-1} \ln N.$$

Hence, using linear interpolation (e.g., see [1, Theorem 3.12]), we conclude that

$$\|\bar{U}^N - u\|_\Omega \le CN^{-1} \ln N,$$

where $\bar{U}^N$ is again the piecewise linear interpolant of the mesh function U^N.

4 Singularly Perturbed Boundary Value Problems with Non-smooth Data

Let us now consider a reaction-diffusion two point boundary value problem with a diffusion coefficient of a constant scale $O(\varepsilon)$ and a lack of smoothness in the data at some internal point. Find $u \in C^1(\bar{\Omega}) \cap C^k(\Omega_d)$ such that

$$-\varepsilon u'' + b(x)u = f(x),\ b(x) \geq \beta > 0,\ x \in \Omega_d,\ u(0), u(1)\,\text{given};\ \ b, f \in C^k(\Omega_d).$$

The associated reduced problem is $b(x)v_0(x) = f(x),\ x \neq d$ and so if $[(f/b)(d)] \neq 0$, then the reduced solution will be discontinuous at the internal point d. Hence, the solution will contain an internal layer function, which will exhibit layers of width $O(\sqrt{\varepsilon}\ln\varepsilon)$ on either side of the interface point $x = d$. Note that by requiring that $u \in C^1(\bar{\Omega})$ we are imposing the constraints $[u](d) = [u'](d) = 0$ on the solution.

We generalize this problem to a reaction-diffusion problem with a variable diffusion coefficient having potentially different scales either side of a point of discontinuity $x = d$ in the data. Find $u \in C^0(\bar{\Omega}) \cap C^4(\Omega_d)$ such that

$$L_{\varepsilon,(3)}u := -(\varepsilon(x)u')' + r(x)u_\varepsilon = f,\ x \in \Omega_d;\quad u(0), u(1)\text{ given}; \tag{5a}$$

$$[f(d)] = Q_2,\ [r(d)] = Q_3,\quad [-\varepsilon u'_\varepsilon(d)] = Q'_1 \leq C(\sqrt{\varepsilon_1} + \sqrt{\varepsilon_2}), \tag{5b}$$

$$\varepsilon(x) = \begin{cases} \varepsilon_1 p(x), x < d \\ \varepsilon_2 p(x), x > d \end{cases},\ \varepsilon_1, \varepsilon_2 > 0,\ p(x) \geq \underline{p} > 0,\quad x \in \Omega_d, \tag{5c}$$

$$p, r, f \in C^4(\Omega_d),\qquad r(x) \geq r_0 > 0,\ \frac{r(x)}{p(x)} > \beta > 0,\quad x \in \Omega_d. \tag{5d}$$

In particular, Eqs. (5c) and (5b) above indicate that all the coefficients in (5a) may exhibit a jump at $x = d$ and also they allow for a scaled jump in the flux $(-\varepsilon u'_\varepsilon)$ at $x = d$. The regular component v and singular component w of the solution are defined, respectively, as the solutions of the discontinuous problems

$$\begin{aligned} L_{\varepsilon,(3)}v = f,\quad L_{\varepsilon,(3)}w = 0,\quad x \in \Omega_d;\ \ r(x)v(x) = f(x),\ x \in \{0, d^-, d^+, 1\}; \\ [w(d)] = -[v(d)],\quad [\varepsilon w'(d)] = -[\varepsilon v'(d)] - Q'_1, \\ w(0) = u(0) - v(0),\ w(1) = u(1) - v(1). \end{aligned}$$

Note that, in general, $v, w \notin C^0(\overline{\Omega})$ even though their sum $u = v + w$ is continuous. For each integer k, satisfying $0 \leq k \leq 4$, these components satisfy the bounds [7].

$$|v_\varepsilon^{(k)}(x)| \leq \begin{cases} C(1 + \varepsilon_1^{1-\frac{k}{2}}),\ x < d \\ C(1 + \varepsilon_2^{1-\frac{k}{2}}),\ x > d \end{cases},$$

$$|w_\varepsilon^{(k)}(x)| \le \begin{cases} C\varepsilon_1^{-\frac{k}{2}}\left(e^{-\sqrt{\frac{\beta}{\varepsilon_1}}x} + e^{-\sqrt{\frac{\beta}{\varepsilon_1}}(d-x)}\right), & x < d \\ C\varepsilon_2^{-\frac{k}{2}}\left(e^{-\sqrt{\frac{\beta}{\varepsilon_2}}(1-x)} + e^{-\sqrt{\frac{\beta}{\varepsilon_2}}(x-d)}\right), & x > d \end{cases}.$$

Based on these bounds, an appropriate piecewise-uniform Shishkin mesh can be constructed. However, to retain parameter-uniform second order convergence for this reaction-diffusion problem, it is necessary to employ a particular discretization of the jump conditions at the mesh point $x_i = d$. See [7] for details.

Boundary and interior layers can be classified as either weak or strong layers. A layer is a strong layer near a point $x = p$ if $u'(p^-)$ or $u'(p^+)$ is unbounded as $\varepsilon \to 0$. A layer is a weak layer near $x = p$, if the first derivatives $u'(p^-)$ and $u'(p^+)$ are bounded, but either $u''(p^-)$ or $u''(p^+)$ is unbounded as $\varepsilon \to 0$. In all of the above problems, only strong interior layers appeared. When a convective term is included in the differential equation, weak interior layers can appear in the solutions. Note that, if one employs a classical finite difference operator (such as simple upwinding), then it is essential that one employs a suitable layer adapted mesh to capture any strong internal layers present in the solution. The adverse effect of using a uniform mesh for a weak layer are minimal. Nevertheless, one still observes some improvement in the numerical results if one also uses a layer-adapted mesh in the vicinity of a weak layer. We refer to the numerical results in [8] to justify this comment.

We now look at five particular singularly perturbed problems, with a convective term present in the differential equation. These particular problems illustrate the variety of layers that can occur when the problem has discontinuous data. For all five problems, we seek to find $u \in C^1(\bar{\Omega})$, with $u(0) = u(1) = 0$ and

$$-\varepsilon u'' + u' = 1,\ x < 0.5; \quad -\varepsilon u'' - u' = -1,\ x > 0.5; \tag{6a}$$
$$-\varepsilon u'' + u' = 1,\ x < 0.5; \quad -\varepsilon u'' + u = -1,\ x > 0.5; \tag{6b}$$
$$-\varepsilon u'' + u' = 1,\ x < 0.5; \quad -\varepsilon u'' + 2u' = -1,\ x > 0.5; \tag{6c}$$
$$-\varepsilon u'' - u' = 1,\ x < 0.5; \quad -\varepsilon u'' + u = -1,\ x > 0.5; \tag{6d}$$
$$-\varepsilon u'' - u' = 1,\ x < 0.5; \quad -\varepsilon u'' + u' = -1,\ x > 0.5. \tag{6e}$$

For the first four problems, we can define the following associated reduced problems

$$v_0' = 1,\ x < 0.5,\ v_0(0) = u(0); \quad v_0' = 1,\ x > 0.5,\ v_0(1) = u(1);$$
$$v_0' = 1,\ x < 0.5,\ v_0(0) = u(0); \quad v_0 = 1,\ x > 0.5;$$
$$v_0' = 1,\ x < 0.5,\ v_0(0) = u(0); \quad 2v_0' = -1,\ x > 0.5,\ [v_0(0.5)] = 0;$$
$$v_0' = 1,\ x < 0.5,\ v_0(0) = u(0); \quad v_0 = -1,\ x > 0.5,\ [v_0(0.5)] = 0.$$

In the case of the first two problems (6a, 6b), the reduced solution is discontinuous and a strong interior layer forms in a neighbourhood of $x = 0.5$. In the next two problem classes (6c, 6d), the reduced solution is continuous and a weak layer forms in a neighbourhood of $x = 0.5$. There is no reduced problem for the fifth problem (6e) as the solution is of order $O(\varepsilon e^{\frac{1}{2\varepsilon}})$ throughout the domain, except in $O(\varepsilon \ln(1/\varepsilon))$-neighbourhoods of the two end points.

For the first four sample problems (6a–6d), associated problem classes can be formulated and parameter-uniform numerical methods (based on standard finite difference schemes combined with appropriately fitted piecewise uniform Shishkin) were constructed in [9]. In the case of problems of the form (6e) a modification of the transmission condition from $[u'(d)] = 0$ to $[(-\varepsilon u' + \gamma u)(d)] = 0$ (where γ sufficiently large) allows one design a parameter-uniform numerical method for this modified class of problems [10].

Further effects can be built into such problem classes, such as point sources (i.e. δ-functions) or multi-parameter problems with variable diffusion. In [11], high order parameter-uniform methods were constructed for the following two problem classes: Find $u \in C^4(\Omega_d) \cap C^0(\bar{\Omega}),\ u(0) = u_0,\ u(1) = u_1$, such that

$$-(\varepsilon(x)u')' + a(x)u' + b(x)u = f(x),\ x \neq d,\quad \big[-(\varepsilon u')(d)\big] = Q_1; \quad (7)$$

$$-(\varepsilon(x)u' + a(x)u)' + b(x)u = f(x),\ x \neq d,\quad \big[(-\varepsilon u' + au)(d)\big] = Q_2; \quad (8)$$

and for both problem classes we assume that

$$\varepsilon, a, b, f \in C^4(\Omega_d);\quad b(x) \geq 0,\quad \varepsilon(x) > 0, |a(x)| > 0,\ x \neq d.$$

The nature of the interior layers appearing in (7) and (8) can have different character. If $Q_1 = 0$ in (7), then the strength of the interior layer depends on the sign of $a(x)$ and on the change in the ratio of convection to diffusion at d. A strong interior layer can appear in (7) when $a(x) > 0, x \neq d$ and

$$\frac{\varepsilon(d^-)}{a(d^-)} << \frac{\varepsilon(d^+)}{a(d^+)}.$$

If $Q_2 = 0$ then a strong interior layer always appears near d in (8).

Numerous different types of interior layers can appear in problem classes (5), (7) and (8), which is an indication of the rich variety of layers one can expect to occur in higher dimensional versions of these one dimensional problem classes.

5 Singularly Perturbed Turning Point Problems

Singularly perturbed differential equations with discontinuous coefficients can be viewed as approximate models for singularly perturbed nonlinear problems. For example, in the case of the quasilinear second order problem

$$-\varepsilon u'' + uu' + u = 0,\quad x \in \Omega; \quad (9a)$$

$$u(0) = A > 1,\qquad u(1) = B < -1; \quad (9b)$$

then an interior layer (with a profile of hyperbolic-tangent type) will appear [12, 13] in the vicinity of some internal point $0 < d_\varepsilon < 1$, where $u(d_\varepsilon) = 0$ and $d_0 := \lim_{\varepsilon\to 0} d_\varepsilon$. The associated reduced problem to this nonlinear problem is the nonlinear first order problem

$$u_0u_0' + u_0 = 0,\ u_0(0) = A,\ 0 < x < d_0;\quad u_0u_0' + u_0 = 0,\ u_0(1) = B,\ 1 > x > d_0;$$

which has the discontinuous solution

$$u_0(x) = A - x,\quad 0 \le x < d_0;\qquad u_0(x) = B + 1 - x,\quad d_0 < x \le 1.$$

For $\varepsilon << 1$, it is natural to consider the following approximate problem for the above nonlinear problem (9). Find $y \in C^1(\Omega)$ such that

$$-\varepsilon y'' + u_0 y' + y = 0,\quad x \in \Omega_d,\quad y(0) = u(0),\ y(1) = u(1).$$

This linearized approximate problem is within the class of problems discussed in Sect. 3, for which parameter-uniform numerical methods have been developed in the literature [9]. However, the convective coefficient in the nonlinear problem is continuous and not discontinuous as in the above linearization of (9).

An alternative linearization of the nonlinear problem (9) would be the following class of turning point problems with a continuous convective coefficient: Find $u \in C^3(\Omega)$ such that

$$(-\varepsilon u'' + a_\varepsilon u' + bu)(x) = f(x),\quad x \in \Omega;\quad u(0) > 0,\ u(1) < 0; \tag{10a}$$
$$a_\varepsilon \in C^2(\Omega),\ a_\varepsilon(x) > 0, x \in [0, d),\quad a_\varepsilon(d) = 0,\quad a_\varepsilon(x) < 0, x \in (d, 1]. \tag{10b}$$

Observe that the convective coefficient is continuous, but depends on the singular perturbation parameter. We also assume that the convective coefficient $a_\varepsilon(x)$ contains it's own interior layer. Define the limiting functions

$$a_0^-(x) := \lim_{\varepsilon\to 0} a_\varepsilon(x),\ x \in [0, d)\ \text{ and }\ a_0^+(x) := \lim_{\varepsilon\to 0} a_\varepsilon(x),\ x \in (d, 1].$$

Assume that

$$|a_\varepsilon(x)| > |\theta \tanh(r(d - x)/\varepsilon)|,\quad \theta > 2r > 0, \tag{10c}$$
$$|(a_0^\pm - a_\varepsilon)(x)| \le Ce^{\pm\frac{\theta}{2\varepsilon}(d-x)},\quad x \in \Omega_d. \tag{10d}$$

Then the solution of the above problem (10) will have an interior layer (with a profile of hyperbolic-tangent type) in an $O(\varepsilon)$-neighbourhood of the point d, where the convective coefficient has an interior layer. Based on this information, a parameter-uniform numerical method was constructed [14] and shown to be parameter-uniformly convergent of first order.

The nature of the interior layers appearing in problem (10) is different to the layers appearing in the solutions of singularly perturbed turning point problems of the form

$$-\varepsilon u'' + a(x)u' + b(x)u = f(x), \quad x \in \Omega;$$
$$a \in C^2(\Omega), \; a(d) = 0, \; d \in \Omega, b > 0;$$

where the convective coefficient a is independent of the singular perturbation parameter. Depending on the quantity $b(d)/a'(d)$, there may be no interior layer or there may be layers of power-law type present at d. See [15, 16] for a discussion of these types of turning point problems.

6 Singularly Perturbed Parabolic Problems

Consider the following singularly perturbed parabolic problem: Find $u \in C^{1+\gamma}(G)$, $G := (0, 1) \times (0, 1]$ such that

$$-\varepsilon u_{ss} + a_1 u_s + b_1 u + c_1 u_t = f, \, (s, t) \in G \setminus \Gamma,$$
$$\Gamma := \{(s, t)|s = d(t), 0 < d(t) < 1\};$$
$$u = g, \, (s, t) \in \bar{G} \setminus G; \quad b(s, t) \geq 0, \; c(s, t) \geq \gamma > 0.$$

As in the previous sections, interior layers can appear in the solution due to discontinuous coefficients a, b, c and/or f [17]. Nine subclasses can be identified (see [9] and [10]), which can exhibit strong or weak interior layers in the vicinity of the curve Γ. Note that in these references, the interior layer location is known and the center of the interior layer can move with time. By using a transformation $T : (s, t) \rightarrow (x, t)$ so that $T : \Gamma \rightarrow \{x = d(0)\}$ any internal layer will be located along the vertical line $x = d$ in the computational domain (x, t) [18]. A computed solution is generated on this transformed domain so that the piecewise-uniform mesh is aligned to the curve Γ. Shishkin [19] established that it is necessary to align the grid to the interior layer if one is seeking to construct a parameter-uniform numerical method.

Moreover, for parabolic problems interior layers can also appear when the boundary/initial conditions are not smooth [20, 21]. If there is a discontinuity in the initial condition $u(s, 0)$ then a standard finite difference operator on a piecewise uniform mesh will not suffice to generate a parameter-uniform numerical method. A special fitted finite difference operator is required [22]. A regularization of a discontinuous initial condition is possible by replacing the initial condition with an initial condition of the form

$$u(s, 0) = A \tanh\left(\frac{s - d_0}{\sqrt{\varepsilon}}\right) + B.$$

In this case, assuming the convective coefficient is independent of space, the initial interior layer is transported along the curve $\{(d(t), t) : d'(t) = a(t), d(0) = d_0\}$; and a parameter-uniform numerical method based on classical finite difference operator, a suitable transformation and an appropriate piecewise-uniform mesh can be constructed [23, 24].

7 Singularly Perturbed Nonlinear Problems with Interior Layers

Semilinear singularly perturbed differential equations of the form

$$-\varepsilon u'' + g(x,u) = 0, \quad x \in \Omega, \quad u(0) = A, \ u(1) = B; \tag{11a}$$

are typically constrained by a condition of the form

$$g_u(x,u) \geq \beta > 0, \quad (x,u) \in \bar{\Omega} \times [-M, M]; \tag{11b}$$

where M is a sufficiently large number that needs to be be explicitly identified. This constraint is a restriction on the admissible type of nonlinear problem being studied. Note that requiring

$$g_u(x,u) \geq \beta > 0, \quad (x,u) \in \bar{\Omega} \times (-\infty, \infty),$$

is a significantly stronger restriction to impose on the problem class. This stronger constraint guarantees a unique solution to the reduced problem and thereby regulates the problem class to a minor extension from the corresponding class of linear problems of reaction-diffusion type

$$-\varepsilon u'' + b(x)u = f(x), \quad b(x) \geq \beta > 0.$$

Interesting new phenomena can be observed when the nonlinear reduced problem $g(x,v) = 0$ has non-unique solutions. The reduced solutions are classified as stable reduced solutions if $g_u(x, v(x)) > 0, \ \forall x \in \bar{\Omega}$ and as unstable reduced solutions if $g_u(x, v(x)) < 0, \ \forall x \in \bar{\Omega}$.

Interior layers can appear in nonlinear problems. Typically, restrictions need to be placed on the data so that solutions to the reduced problem exist and for the solution of the singularly perturbed problem to exist and be unique. For example, in the case of the semilinear reaction-diffusion problem: Find $u \in C^1(\bar{\Omega}) \cap C^3(\Omega_d)$ such that

$$-\varepsilon u'' + (1-u^2)u = f(x), x \in \Omega_d; \qquad d = 0.5;$$
$$u(0) = A, u(1) = B; \quad f(x) > 0, x < d, \ f(x) < 0, x > d; \ [f](d) \neq 0;$$

we impose the following limits on the input data

$$|A|,\ |B| < \frac{1}{\sqrt{3}},\quad \|f\| < \frac{2}{3\sqrt{3}}.$$

This problem is formulated so that a discontinuous stable reduced solution lies between two discontinuous unstable reduced solutions. Interior layers can appear in the solution of this problem and the location of the layer will be positioned around the point d, where the discontinuity in the data is located. By placing further restrictions on the data, a parameter-uniform method was constructed in [25] for this semilinear problem.

However, other semilinear problems of the form (11) can be very difficult to solve numerically. In [26] a semilinear problem of the form (11) with smooth data, where an unstable continuous reduced solution was positioned between two stable continuous reduced solutions, was examined. Using a piecewise-uniform Shishkin mesh (of an appropriate width) centered at *any point* in the domain Ω, then an interior layer forms within the fine mesh, no matter where the mesh is centered [26]. Only in the exceptional case where the fine mesh is located in an $O(\sqrt{\varepsilon})$ neighbourhood of the actual location of the interior layer will the numerical approximation be of any true value.

Parameter-uniform numerical methods (based on piecewise-uniform Shishkin meshes) have also been constructed [27] for quasilinear problems with interior layers of the form: Find $u \in C^1(\bar{\Omega}) \cap C^3(\Omega_d)$ such that

$$\varepsilon u''(x) + b(x,u)u'(x) = f(x),\ x \in \Omega_d, \qquad u(0) = A,\quad u(1) = B, \tag{12a}$$

$$b(x,u) = \begin{cases} b_1(u) = -1 + cu, & x < d \\ b_2(u) = 1 + cu, & x > d \end{cases},\quad f(x) = \begin{cases} -\delta_1 < 0, & x < d \\ \delta_2 > 0, & x > d \end{cases} \tag{12b}$$

$$-1 < u(0) < 0,\quad 0 < u(1) < 1,\quad 0 < c \leq 1. \tag{12c}$$

As in the case of the semilinear problem, additional constraints need to be imposed on the data $\{A, B, \|f\|, c\}$ in order for the theoretical convergence result given in [27] to apply. The numerical results in [28] suggest that the numerical approximations generated by the method described in [27] converge for a wider class of problems to that covered by the theoretical convergence analysis in [27]. Note, again, that for this problem (12) the location of the interior layer is known to be positioned at d, where both the convective coefficient $b(x,u)$ and the forcing term f are formulated to be discontinuous.

An interesting open issue is to examine singularly perturbed problems with an interior layer, whose location is not known a priori. Many nonlinear singularly perturbed problems of interest [29–32] exhibit this phenomenon. The design of parameter-uniform numerical methods for a broad class of nonlinear singularly perturbed problems with interior layers, remains an area with significant challenges for the numerical analyst.

References

1. P.A. Farrell, A.F. Hegarty, J.J.H. Miller, E. O'Riordan, G.I. Shishkin, *Robust Computational Techniques for Boundary Layers* (CRC Press, Boca Raton, 2000)
2. G.I. Shishkin, L.P. Shishkina, *Discrete Approximation of Singular Perturbation Problems* (Chapman and Hall/CRC Press, 2008)
3. J.J.H. Miller, E. O'Riordan, G.I. Shishkin, *Fitted Numerical Methods for Singular Perturbation Problems* (World-Scientific, Singapore, 2012) Revised edition
4. V.B. Andreev, Estimating the smoothness of the regular component of the solution to a one-dimensional singularly perturbed convection-diffusion equation. Comput. Math. Math. Phys. **55**(1), 19–30 (2015)
5. G.I. Shishkin, Grid approximation of singularly perturbed parabolic reaction-diffusion equations with piecewise smooth initial-boundary conditions. Math. Model. Anal. **12**(2), 235–254 (2007)
6. P.A. Farrell, A.F. Hegarty, J.J.H. Miller, E. O'Riordan, G.I. Shishkin, Global maximum norm parameter-uniform numerical method for a singularly perturbed convection-diffusion problem with discontinuous convection coefficient. Math. Comput. Model. **40**, 1375–1392 (2004)
7. C. de Falco, E. O'Riordan, Interior layers in a reaction–diffusion equation with a discontinuous diffusion coefficient. Int. J. Numer. Anal. Model. **7**(3), 444–461 (2010)
8. P.A. Farrell, A.F. Hegarty, J.J.H. Miller, E. O'Riordan, G.I. Shishkin, Singularly perturbed convection diffusion problems with boundary and weak interior layers. J. Comp. Appl. Maths. **166** (1), 133–151 (2004)
9. R.K. Dunne, E. O'Riordan, *Interior layers arising in linear singularly perturbed differential equations with discontinuous coefficients*, eds. by I. Farago, P. Vabishchevich, L. Vulkov. Proceedings Fourth International Conference on Finite Difference Methods (Rousse University, Bulgaria, 2007), pp. 29–38
10. E. O'Riordan, Opposing flows in a one dimensional convection-diffusion problem, Central Eur. J. Math. **10**(1), 85–100 (2012)
11. C. de Falco, E. O'Riordan, A parameter robust Petrov-Galerkin scheme for advection-diffusion-reaction equations. Numer. Algor. **56**(1), 107–127 (2011)
12. N.N. Nefedov, L. Recke, K.R. Schneider, Existence and asymptotic stability of periodic solutions with an interior layer of reaction-advection-diffusion equations. J. Math. Anal. Appl. **405**, 90–103 (2013)
13. G.I. Shishkin, Difference approximations of the Dirichlet problem for a singularly perturbed quasilinear parabolic equation in the presence of a transition layer. Russian Acad. Sci. Dokl. Math. **48**(2), 346–352 (1994)
14. E. O'Riordan, J. Quinn, A singularly perturbed convection diffusion turning point problem with an interior layer. Comput. Meth. Appl. Math. **12**(2), 206–220 (2012)
15. A.E. Berger, H. Han, R.B. Kellogg, A priori estimates and analysis of a numerical method for a turning point problem. Math. Comp. **42**(166), 465–492 (1984)
16. P.A. Farrell, Sufficient conditions for the uniform convergence of a difference scheme for a singularly perturbed turning point problem. SIAM J. Numer. Anal. **25**(3), 618–643 (1988)
17. G.I. Shishkin, A difference scheme for a singularly perturbed parabolic equation with discontinuous coefficients and concentrated factors, U.S.S.R. Comput. Math. Math. Phys. **29**, 9–19 (1989)
18. E. O'Riordan, G.I. Shishkin, Singularly perturbed parabolic problems with non-smooth data. J. Comp. Appl. Maths. **166**(1), 233–245 (2004)
19. G.I. Shishkin, Limitations of adaptive mesh refinement techniques for singularly perturbed problems with a moving interior layer. J. Comput. Appl. Math. **166**(1), 267–280 (2004)
20. G.I. Shishkin, A difference scheme for a singularly perturbed equation of parabolic type with a discontinuous initial condition. Soviet Math. Dokl. **37**, 792–796 (1988)
21. G.I. Shishkin, A difference scheme for a singularly perturbed equation of parabolic type with a discontinuous boundary condition, U.S.S.R. Comput. Math. Math. Phys. **28**, 32–41 (1988)

22. P.W. Hemker, G.I. Shishkin, Approximation of parabolic PDEs with a discontinuous initial condition. East-West J. Numer. Math **1**, 287–302 (1993)
23. J.L. Gracia, E. O'Riordan, A singularly perturbed convection–diffusion problem with a moving interior layer. Int. J. Numer. Anal. Model. **9**(4), 823–843 (2012)
24. J.L. Gracia, E. O'Riordan, A singularly perturbed parabolic problem with a layer in the initial condition. Appl. Math. Comput. **219**(2), 498–510 (2012)
25. P.A. Farrell, E. O'Riordan, G.I. Shishkin, A class of singularly perturbed semilinear differential equations with interior layers. Math. Comp. **74**, 1759–1776 (2005)
26. N. Kopteva, M. Stynes, Stabilised approximation of interior-layer solutions of a singularly perturbed semilinear reaction-diffusion problem. Numer. Math. **119**, 787–810 (2011)
27. P.A. Farrell, E. O'Riordan, G.I. Shishkin, A class of singularly perturbed quasilinear differential equations with interior layers. Math. Comp. **78**(265), 103–127 (2009)
28. P.A. Farrell, E. O'Riordan, *Examination of the performance of robust numerical methods for singularly perturbed quasilinear problems with interior layers*, eds. by A. Hegarty, N. Kopteva, E. O'Riordan, M. Stynes. Proceedings BAIL, Lecture Notes in Computational Science and Engineering Springer 2009, vol. 69, pp. 141–152 (2008)
29. K.W. Chang, F.A. Howes, *Nonlinear Singular Perturbation Phenomena* (Springer-Verlag, New York, 1984)
30. F.A. Howes, Boundary-interior layer interactions in nonlinear singular perturbation theory. Mem. AMS, **15** (203), (1978)
31. A.M. Ilin, *Matching of Asymptotic Expansions of Solutions of Boundary Value Problems*, Transactions of Mathematical Monographs, vol. 102 (American Mathematical Society, 1991)
32. R.E. O'Malley, *Singular Perturbation Methods for Ordinary Differential Equations* (Springer, New York, 1991)

Singularly Perturbed Delay Differential Equations and Numerical Methods

Ramanujam Narasimhan

Abstract The main objective of my talk is to discuss some numerical methods for singularly perturbed delay differential equations. First some well-known mathematical models represented by differential equations with out delay and with delay are presented. Then some basic numerical methods for delay differential equations are briefly described. After this an introduction to singularly perturbed delay problems is given. Finally some numerical methods for these problems are discussed.

Keywords Singular perturbation problems · Delay differential equations · Finite difference methods

1 Introduction

As mentioned in the abstract first I will present some well known mathematical models represented by differential equations. Let $P(t)$ denote the population size of a single species at time t. Let b and d denote birth and death rates respectively. Then a simple population model is [1]

$$\frac{dP}{dt} = bP - dP = rP,$$

where $r = b - d$ is the intrinsic growth rate of the population. This model is valid, in general, for short periods. Taking into account that resources are limited then the more realistic model will be

$$\frac{dP}{dt} = rP\left(1 - \frac{P}{K}\right),$$

R. Narasimhan (✉)
Department of Mathematics, School of Mathematical Sciences,
Bharathidasan University, Tiruchirappalli 620024, Tamil Nadu, India
e-mail: matram2k3@gmail.com

V. Sigamani et al. (eds.), *Differential Equations and Numerical Analysis*,
Springer Proceedings in Mathematics & Statistics 172,
DOI 10.1007/978-81-322-3598-9_3

$r > 0,\ K > 0$ are respectively intrinsic growth and carrying capacity. Next a prey-predator model is described. Let $x(t),\ y(t)$ denote the populations of the prey and predator species at time t respectively. Under suitable assumptions this model can be described by the following system [1, 2]

$$\begin{cases} \frac{dx}{dt} = ax - bxy = x(a - by),\ a, b > 0, \quad x(0) = x_0, \\ \frac{dy}{dt} = -py + qxy = -y(p - qx),\ p, q > 0, \quad y(0) = y_0. \end{cases}$$

In many applications, one assumes that the future state of the system is independent of the past states and is determined solely by the present. However, under closer scrutiny, it becomes apparent that this is often a first order approximation to the true situation and more realistic model would involve some of the past states of the system. For an example

$$\begin{cases} \frac{dx}{dt} = g(x(t), x(s), t), \quad t - \tau \le s < t, \\ x(0) = C. \end{cases}$$

Now I present a simple population delay model. Imagine a biological population composed of adult and juvenile individuals. Let $A(t)$ denote the density of adults at time t. We assume that

- The length of the juvenile period is exactly h units of time for each individual.
- Adults produce offspring at a per capita rate α and that their probability per unit of time of dying is μ.
- A newborn survives the juvenile period with probability ρ.

Let $r = \alpha\rho$. Then we have

$$\frac{dA}{dt} = -\mu A(t) + rA(t - h),$$

where the term $rA(t - h)$ means that newborns become adults with some delay.

In the above discussed logistic model it was assumed that the growth rate of a population at any time t depends on the relative number of individuals at that time. In practice, the process of reproduction is not instantaneous. Hence the more realistic logistic model will be

$$\frac{dP}{dt} = aP(t)\left[1 - \frac{P(t - \tau)}{N}\right],$$

where τ is a delay or lag and the equation is known as Hutchinson's equation or delayed logistic equation.

Let me present a population model consisting of adult and juvenile. Let $A(t)$ and $J(t)$ denote adult and juvenile population respectively. Further let 10 be the dividing line for sexually matured. Then the mathematical model for this population problem is

$$\begin{cases} \frac{dJ}{dt} = bA(t) - s_a bA(t-10) - d_j J(t), \\ \frac{dA}{dt} = sbA(t-10) - d_a A(t), \\ J(0) = J_0, \quad A(t) = A_0, \quad -10 \le t \le 0. \end{cases}$$

Next let us consider a cell to cell virus spread model. Interaction between the susceptible host cells (S), infected host cells (I) and free virus particles (V) can be modelled as

$$\begin{cases} \frac{dS}{dt} = \lambda S(t) - dS(t) - \beta S(t)V(t), \\ \frac{dI}{dt} = \beta S(t)V(t) - aI(t), \\ \frac{dV}{dt} = kI(t) - \beta S(t)V(t) - uV(t), \end{cases}$$

where $\lambda S(t), dS(t), \beta S(t)V(t)$ are respectively production, death and infected rates of the susceptible host cells S, etc. In above model it was assumed that as soon as the virus contacts a target cell, the cell begins producing virus. In general this may not happen. In fact there is a time delay between the initial viral entry into a cell and subsequent viral production. Therefore an appropriate model will be [3]

$$\begin{cases} \frac{dS}{dt} = \lambda S(t) - dS(t) - \beta S(t)V(t), \\ \frac{dI}{dt} = \frac{\beta S(t-\tau)V(t-\tau)}{1+V(t-\tau)} - aI(t), \\ \frac{dV}{dt} = kI(t) - \beta S(t)V(t) - uV(t). \end{cases}$$

2 Preliminaries

Definition 1 *(Functional Differential Equation (FDE))* A FDE is an equation for an unknown function which involves derivatives of the function and in which the function, and possibly its derivatives, occur with various different arguments.

Examples:

- $u'(x) = u(x - \pi)$ (DDE),
- $u'(x) = u(x) - u(x/2)$ (DDE with variable delay),
- $u'(x) = x^2 u(x) - u'(x-1)$ (NFDE),
- $u'(x) = u(x)u(x-1) + u(x+2)$ (DDE with advance argument),
- $u''(x) = -u'(x) + \sin(u(x)) + u(x-5) + u^2(x-3)$ (non-linear DDE),
- $u'(x) = u^2(x) + \frac{2}{\pi}\int_0^\infty e^{-s^2} u(x-s)ds$ (DDE with distribution delay).

Definition 2 *(Delay Differential Equation (DDE))* A retarded functional DE or a DE with retarded arguments or a DDE is a FDE when the highest derivative only occurs with one value of the arguments, and this argument is not less than the arguments of the unknown function and its lower order derivatives appearing in the equation.

2.1 Properties of FDE

- **Propagation of discontinuity**:

Consider the DDE

$$\begin{cases} u'(x) = -u(x-1), & x \geq 0, \\ u(x) = 1, & x \leq 0. \end{cases}$$

Then we see that $u \in C^1(0,1] \cap C^0[0,1]$, u' is discontinuous at $x = 0$ and u'' is discontinuous at $x = 1$.

- **No injectivity**:

Consider the DDE

$$u'(x) = u(x-1)(u(x)-1), \quad t \geq 0.$$

Then $u(x) = 1$ is the solution for any initial function $\phi(x)$ on $[-1, 0]$ such that $\phi(0) = 1$.

- **Propagation of discontinuity**:

Consider the FDE

$$\begin{cases} u'(x) = -u'(x-1), & t \geq 0, \\ u(x) = x, & x \leq 0. \end{cases}$$

Then $u'(t)$ has a jump discontinuity at $x = 0, x = 1, \cdots$.

- **Non uniqueness**:

Consider the DDE

$$\begin{cases} u'(x) = u(x - |u(x)| - 1) + \frac{1}{2}, & x \geq 0, \\ u(x) = \phi(x), & x \leq 0. \end{cases}$$

where

$$\phi(x) = \begin{cases} 1, & x < -1, \\ 0, & -1 \leq x \leq 0. \end{cases}$$

On $[0, 2]$, $u(x) = \frac{3}{2}x$ and $u(x) = \frac{x}{2}$ are solutions.

2.2 Classification of FDEs

- **Retarded type**:
  (delay will not occur in the highest derivative)

$$\frac{du}{dx} + au(x) + bu(x - \tau(x)) = g(x).$$

- **Neutral type**:
  (delay will appear in the highest derivative)

$$\frac{du}{dx} + a\frac{du(x - \tau(x))}{dx} + bu(x) + cu(x - \tau(x)) = g(x).$$

- **Advanced type**:
  (delay will appear in the highest derivative and not in the next lower order)

$$\frac{d^2u(x - \tau(x))}{dx^2} + a\frac{du}{dx} + bu(x) + cu(x - \tau(x)) = g(x).$$

- **State dependent type**:
  (delay can be a function of unknown function)

$$\frac{du}{dx} + au(x) + bu(x - \tau(x, u(x))) = g(x).$$

2.3 Method of Steps/Step by Step Integration

2.3.1 Initial Value Problem (IVP)

Consider the following IVP for the first order ordinary delay differential equation(ODDE):

$$\begin{cases} u' + au(x) + bu(x - \tau) = g(x), & x > 0, \\ u(x) = \phi(x), & x \in [-\tau, 0]. \end{cases} \tag{1}$$

Recall that $\phi(x)$ is the **history function**. The solution of the above problem is obtained as follows:

$$\begin{cases} u_A' + au_A(x) = -b\phi(x - \tau) + g(x), & x \in (0, \tau], \\ y_A(0) = \phi(0), \\ u_B' + au_B(x) = -bu_A(x - \tau) + g(x), & x \in (\tau, 2\tau], \\ u_B(\tau) = u_A(\tau) \end{cases}$$

and so on.

2.3.2 Boundary Value Problem (BVP)

Consider the following BVP for the second order ODDE:

$$\begin{cases} u''(x) + au'(x) + bu(x) + cu(x-1) = f(x), & x \in (0,2), \\ u(x) = \phi(x), & x \in [-1,0], \\ u(2) = l. \end{cases} \tag{2}$$

This BVP can be solved as follows:

$$\begin{cases} u_A''(x) + au_A'(x) + bu_A(x) = f(x) - c\phi(x-1), & x \in (0,1), \\ u_B''(x) + au_B'(x) + bu_B(x) + cu_A(x-1) = f(x), & x \in (1,2), \\ u_A(0) = \phi(0), \quad u_B(2) = l, \\ u_A(1-) = u_B(1+), \quad u_A'(1-) = u_B'(1+). \end{cases}$$

2.4 *Existence, Uniqueness and Stability*

Consider the initial value problem

$$\begin{cases} u' = f(x, u(x), u(x-\tau)), & x > 0, \\ u(x) = \phi(x), & x \in [-\tau, 0]. \end{cases} \tag{3}$$

On $[0, \tau]$, this IVP becomes

$$\begin{cases} u' = f(x, u(x), \phi(x-\tau)), & x > 0, \\ u(0) = \phi(0), \end{cases}$$

which is an IVP for ODE with out a delay term and hence it can be solved. On $[\tau, 2\tau]$, the term $u(x-\tau)$ in f is known and the initial value $u(\tau)$ is also known. Repetition shows the existence, uniqueness and continuous dependence on the data of the solution for all $x > 0$.

3 Euler Method for Delay Differential Equations

One cannot apply directly the existing numerical method of ODE to DDEs. To illustrate this, we consider the following IVP:

$$\begin{cases} u'(x) = u(x-\tau), & x > 0, \tau > 0, \\ u(x) = 1, & x \in [-\tau, 0]. \end{cases} \tag{4}$$

The Euler formula for the above problem is given by

$$\begin{cases} u(x_i) = u(x_{i-1}) + h(i)(u(x_{i-1} - \tau)), \\ u(x) = 1, \quad x \in [-\tau, 0]. \end{cases} \tag{5}$$

Note that, we used the Taylor's expansion to discretize the DDE into difference equation [4, 5]. But the solution is no longer smooth in the domain of differential equation. It is smooth except at the points of the discontinuity ($x = 0,\ 1$). Since the truncation error depends on the higher derivatives of the solution we have to

- include all the points of the discontinuity as mesh points,
- go for a non uniform mesh,
- apply an appropriate interpolation method to evaluate at the points $x_i - \tau$ as they need not fall on the mesh points.

Hence the the appropriate Euler method is

$$\begin{cases} u(x_i) = u(x_{i-1}) + h(i)(u^I(x_i)), & i \geq 1, \\ u(x_0) = 1, \end{cases} \tag{6}$$

where

$$u^I(x_i) = \begin{cases} u_0(x_i - 1), & x_i \leq 1, \\ u(x_j), & x_i - 1 = x_j,\ x_i > 1, \\ \frac{(x_i-1)-x_j}{x_{j+1}-x_j} u(x_{j+1}) + \frac{x_{j+1}-(x_i-1)}{x_{j+1}-x_j} u(x_j), & x_j < x_i - 1 < x_{j+1},\ x_i > 1. \end{cases} \tag{7}$$

This method is known as continuous ODE method [6]. The above numerical method is consistent with the DDE (4). (Local truncation error tends to zero as the mesh parameter tends to zero.)

4 Singular Perturbation Problems

Definition 3 Let P_ε denote the original problem and u_ε be its solution. Let P_0 denote the reduced problem of P_ε (setting $\varepsilon = 0$ in P_ε). Then the problem P_ε is called a **Singular Perturbation Problem (SPP)** if and only if u_ε does not converge uniformly to u_0 in the entire domain of the definition of the problem. Otherwise the problem is called **Regular Perturbation Problem (RPP)** [7–9]

4.1 One-Dimension Convection Diffusion Problems

Consider the following differential equation

$$-\varepsilon u''(x) + a(x)u'(x) + b(x)u(x) = f(x), \quad x \in \Omega, \tag{8}$$

subject to

- Dirichlet type boundary conditions

$$u(0) = A,\ u(1) = B.$$

- Neumann type boundary conditions

$$u'(0) = A,\ u'(1) = B.$$

- Mixed type boundary conditions

$$\begin{cases} \alpha_1 u(0) - \alpha_2 u'(0) = A, \\ \beta_1 u(1) + \beta_2 u'(1) = B. \end{cases}$$

4.2 Reaction Diffusion Problems

Consider the following differential equation

$$-\varepsilon u''(x) + b(x)u(x) = f(x), \quad x \in \Omega, \tag{9}$$

subject to

- Dirichlet type boundary condition

$$u(0) = A,\ u(1) = B.$$

- Neumann type boundary condition

$$u'(0) = A,\ u'(1) = B.$$

- Mixed type boundary condition

$$\begin{cases} \alpha_1 u(0) - \alpha_2 u'(0) = A, \\ \beta_1 u(1) + \beta_2 u'(1) = B. \end{cases}$$

5 Locations of Boundary Layers

Consider the BVP

$$\begin{cases} -\varepsilon u''(x) + a(x)u'(x) + b(x)u(x) = f(x), & x \in \Omega = (0, 1), \\ u(0) = u_0, \quad u(1) = u_1. \end{cases}$$

1. Case (A): $a,\ b,$ and f are smooth.
   - $a(x) > 0,\ \forall x \in \overline{\Omega}$ there is a strong boundary layer at $x = 1$.
   - $a(x) < 0,\ \forall x \in \overline{\Omega}$ there is a strong boundary layer at $x = 0$.
2. Case (B): a and b are smooth, f is bounded and has a discontinuity at $x = d \in \Omega$.
   - $a(x) > 0,\ \forall x \in \overline{\Omega}$ there is a strong boundary layer at $x = 1$ and a weak interior layer at $x = d$ (left side of the point $x = d$).
   - $a(x) < 0,\ \forall x \in \overline{\Omega}$ there is a strong boundary layer at $x = 0$ and a weak interior layer at $x = d$ (right side of the point $x = d$).
3. Case (C): b is smooth, a and f are bounded and have a discontinuity at $x = d \in \Omega$.
   - $a(x) > 0,\ \forall x \in \overline{\Omega}$ there is a strong boundary layer at $x = 1$ and a weak interior layer at $x = d$ (left side of the point $x = d$).
   - $a(x) < 0,\ \forall x \in \overline{\Omega}$ there is a strong boundary layer at $x = 0$ and a weak interior layer at $x = d$ (right side of the point $x = d$).
   - $a(x) > 0,\ x \in (0, d)$ and $a(x) < 0,\ x \in (d, 1)$ there are strong twin interior layers at $x = d$.
   - $a(x) < 0,\ x \in (0, d)$ and $a(x) > 0,\ x \in (d, 1)$ the solution is unbounded.
4. Case (D): Suppose $a(x) = 0$ (reaction diffusion) and
   - the coefficients are all smooth then, there are strong boundary layers at $x = 0$ and $x = 1$.
   - f is discontinuous at $x = d$ then, there are strong boundary layers at $x = 0,\ x = 1,\ x = d$ (both sides).

6 Numerical Methods for SPPs

In general classical numerical methods on equidistant grids yield satisfactory numerical solution for singularly perturbed boundary value problems only if one uses an unacceptably large number of grid points. Further Galerkin finite element method even on layer adapted meshes produces an oscillation of the solution/ and its derivative. Hence one has to develop special types of numerical methods to SPPs. In the literature various non-classical methods are available [10–17]:

(i) Variable Mesh size Method (VMM)
(ii) Boundary Value Technique (BVT)
(iii) Initial Value Technique (IVT)
(iv) Fitted Operator Method (FOM)
(v) Fitted Mesh Method (FMM)
(vi) Booster Method (BM)
(vii) Schwartz Iterative Method (SIM)
(viii) Shooting Method (SM)
(ix) Spline Approximation Method (SAM)
(x) Finite Element Method (FEM)
(xi) Asymptotic Numerical Method (ANM)
(xii) Collocation Method

6.1 Fitted Mesh Method for Second Order SPDEs [18, 19]

Consider the following BVP. Find $u \in Y = C^0(\overline{\Omega}) \cap C^2(\Omega)$ such that

$$\begin{cases} -\varepsilon u''(x) + a(x)u'(x) = f(x), & x \in \Omega = (0,1), \\ u(0) = u_0, \; u(1) = u_1, \end{cases} \tag{10}$$

where u_0, u_1 are given constants, the functions a and f sufficiently smooth functions. Further, we assume that, $a(x) \geq \alpha > 0$. Since the BVP exhibits a boundary layer at $x = 1$, we choose a piecewise uniform mesh on $[0, 1]$. For this we divide the interval $[0, 1]$ into two subintervals, namely $\Omega_1 = [0, 1 - \tau]$, $[1 - \tau, 1]$, where $\tau = \min\{0.5, 2\varepsilon \ln N/\alpha\}$. Let $h = 2N^{-1}\tau$, $H = 2N^{-1}(1 - \tau)$. The fitted mesh $\overline{\Omega}^N = \{x_i\}_{i=0}^N$ is defined by

$$\begin{cases} x_0 = 0, \\ x_i = i * H, & i = 1(1)N/2, \\ x_{N/2+i} = x_{N/2} + i * h, & i = 1(1)N/2. \end{cases}$$

The fitted mesh method for the above problem is given by

$$\begin{cases} L^N U_i = -\varepsilon\delta^2 U_i + a(x_i)D^- U_i = f(x_i), \; i = 1(1)N - 1, \\ U_0 = u_0, \; U_N = u_1, \end{cases}$$

where δ^2, D^- are central and backward difference operators.

The above system yields a numerical solution for the BVP (30).

7 Numerical Method for Singularly Perturbed Delay Differential Equations (SPPDDs) of First Order

Consider the IVP:
Find $u \in C([0, 2]) \cap C^1((0, 2])$ such that

$$\begin{cases} \varepsilon u' + au(x) + bu(x-1) = f(x), & x \in (0, 2], \\ u(x) = \phi(x), & x \in [-1, 0], \end{cases} \tag{11}$$

where $a > 0$. The equivalent problem is given by

$$\begin{cases} \varepsilon u'_A + au_A(x) = f(x) - b\phi(x-1), & x \in (0, 1], \\ u_A(0) = \phi(0), \\ \varepsilon u'_B + au_B(x) = f(x) - bu_A(x-1), & x \in (1, 2], \\ u_B(1) = u_A(1). \end{cases}$$

The problem (11) has a boundary layer at $x = 0$ and an interior layer (due to the delay term) at $x = 1$. So we divide the given domain $[0, 2]$ into four subintervals $[0, \tau]$, $[\tau, 1]$, $[1, 1+\tau]$ and $[1+\tau, 2]$ where $\tau = \frac{2\varepsilon \ln N}{a}$ is transition parameter. On each subinterval there are $\frac{N}{4}$ mesh points are placed. The Shishkin mesh Ω^N is given by

$$\Omega^N = \{0 = x_0, \cdots, x_{\frac{N}{2}} = 1, \cdots, x_N = 2\}.$$

We now define a fitted mesh method on the mesh Ω^N as

$$u_{i+1} = \begin{cases} u_i + \frac{h_i}{\varepsilon}\left[f(x_i) - au_i - b\phi(x_i - 1)\right], & i = 1(1)\frac{N}{2}, \\ u_i + \frac{h_i}{\varepsilon}\left[f(x_i) - au_i - bu_{i-\frac{N}{2}}\right], & i > \frac{N}{2}. \end{cases} \tag{12}$$

This gives a numerical solution for the IVP (11)

8 Location of Boundary Layers for Second Order SPPDDs

Consider the following BVP for SPDDE:

$$\begin{cases} -\varepsilon u''(x) + a(x)u'(x) + b(x)u(x) + c(x)u(x-1) = f(x), & x \in \Omega = (0, 2), \\ u(x) = \phi(x), & x \in [-1, 0], \\ u(2) = u_2. \end{cases}$$

1. Case (A): $a,\ b,\ c,\ f,$ and ϕ are smooth.
   - $a(x) > 0,\ \forall x \in \overline{\Omega}$ there is a strong boundary layer at $x = 2$.
   - $a(x) < 0,\ \forall x \in \overline{\Omega}$ there is a strong boundary layer at $x = 0$ and weak interior layer at $x = 1$ (right side of the point $x = 1$).
2. Case (B): a and b are smooth, f is bounded and has a discontinuity at $x = 1$.
   - $a(x) > 0,\ \forall x \in \overline{\Omega}$ there is a strong boundary layer at $x = 2$ and a weak interior layer at $x = 1$ (left side of the point $x = 1$).
   - $a(x) < 0,\ \forall x \in \overline{\Omega}$ there is a strong boundary layer at $x = 0$ and a weak interior layer at $x = 1$ (right side of the point $x = 1$).
3. Case (C): b is smooth, a and f are bounded and have a discontinuity at $x = 1 \in \Omega$.
   - $a(x) > 0,\ \forall x \in \overline{\Omega}$ there is a strong boundary layer at $x = 2$ and a weak interior layer at $x = 1$ (left side of the point $x = 1$).
   - $a(x) < 0,\ \forall x \in \overline{\Omega}$ there is a strong boundary layer at $x = 0$ and a weak interior layer at $x = 1$ (right side of the point $x = 1$).
   - $a(x) > 0,\ x \in (0, 1)$ and $a(x) < 0,\ x \in (1, 2)$ there is a strong twin interior layer at $x = 1$ and weak interior layer at $x = 2$.
   - $a(x) < 0,\ x \in (0, 1)$ and $a(x) > 0,\ x \in (1, 2)$ the solution is unbounded.
4. Case (D): Suppose $a(x) = 0$ (reaction diffusion) there are strong boundary layers at $x = 0,\ x = 2,$ and $x = 1$ (both sides).

9 Numerical Methods for Singularly Perturbed Second Order Delay Differential Equations

Consider the following BVP:
Find $u \in Y = C^0(\overline{\Omega}) \cap C^2(\Omega)$ such that

$$\begin{cases} -\varepsilon u''(x) + a(x)u'(x) + b(x)u(x) + c(x)u(x-1) = f(x), & x \in \Omega, \\ u(x) = \phi(x), \quad x \in [-1, 0], \\ u(2) = l. \end{cases} \tag{13}$$

where

- $a(x) \geq \alpha > 0, b(x) \geq \beta_0, \gamma_0 \leq c(x) \leq \gamma < 0, 2\alpha + 5\beta_0 + 5\gamma_0 \geq \eta > 0$
- $a,\ b,\ c,\ f,$ and ϕ are sufficiently smooth functions on $\overline{\Omega}$.

The above problem is equivalent to

$$\begin{cases} L_\varepsilon u(x) := \begin{cases} -\varepsilon u''(x) + a(x)u'(x) + b(x)u(x) = f(x) - c(x)\phi(x-1), \ x \in \Omega^-, \\ -\varepsilon u''(x) + a(x)u'(x) + b(x)u(x) + c(x)u(x-1) = f(x), \ x \in \Omega^+, \end{cases} \\ u(0) = \phi(0),\ u(1-) = u(1+),\ u'(1-) = u'(1+),\ u(2) = l. \end{cases} \tag{14}$$

The differential-difference operator L_ε defined above satisfies the following **maximum principle** and in turn yields a **stability result**.

Theorem 1 (Mamimum Principle) *Let* $w \in Y'$ *be any function satisfying* $w(0) \geq 0$, $w(2) \geq 0$, $L_\varepsilon w(x) \geq 0$, $\forall x \in \Omega^- \cup \Omega^+$ *and* $w'(1+) - w'(1-) = [w'](1) \leq 0$. *Then* $w(x) \geq 0$, $\forall x \in \overline{\Omega}$.

Corollary 1 (Stability Result) *If* $u \in Y'$ *then*

$$| u(x) | \leq C \max\left\{ | u(0) |, | u(2) |, \max_{x\in\Omega^-\cup\Omega^+} | L_\varepsilon u(x) | \right\}, \quad \forall x \in \overline{\Omega}.$$

9.1 Initial Value Technique [20]

9.1.1 An Asymptotic Expansion

Let $u_0 \in C^0(\overline{\Omega}) \cap C^1(\Omega \cup \{2\})$ be the solution of the reduced problem of (13) given by

$$\begin{cases} a(x)u_0'(x) + b(x)u_0(x) + c(x)u_0(x-1) = f(x), & x \in \Omega \cup \{2\}, \\ u_0(x) = \phi(x), & x \in [-1, 0], \end{cases} \tag{15}$$

and assume that $\| u_0^{(2)} \| \leq C$. Further, let $v_r(x) = \exp(-\int_x^2 \frac{a(s)}{\varepsilon} ds)$, $\forall x \in \overline{\Omega}$ be the solution of the terminal value problem (TVP)

$$\begin{cases} \varepsilon v_r'(x) - a(x)v_r(x) = 0, & x \in [0, 2), \\ v_r(2) = 1. \end{cases} \tag{16}$$

An asymptotic expansion solution of the original problem (13) is given by

$$u_{as}(x) = \begin{cases} u_0(x) + k_1 x, & x \in [0, 1], \\ u_0(x) + k_2 v_r(x) + k_3, & x \in [1, 2], \end{cases} \tag{17}$$

where the constants k_1, k_2, and k_3 are to be determined such that $u_{as} \in Y'$. In fact the constants k_1, k_2, and k_3 are given by

$$\begin{cases} k_2 = \frac{l - u_0(2)}{1 + (\frac{a(1)}{\varepsilon} - 1)v_r(1)}, \\ k_3 = k_2(\frac{a(1)}{\varepsilon} - 1)v_r(1), \text{ and} \\ k_1 = k_2(\frac{a(1)}{\varepsilon})v_r(1). \end{cases} \tag{18}$$

Theorem 2 *The function u_{as} defined by (17) satisfies the inequality*

$$| u(x) - u_{as}(x) | \le C\varepsilon, \quad x \in \overline{\Omega}.$$

Here $u(x)$ is the solution of the problem (13).

9.1.2 Numerical Method

Since the BVP (13) exhibits a boundary layer at $x = 2$ and $u_0''(x)$ has jump discontinuity at $x = 1$ we choose a piece-wise uniform Shishkin mesh on [0, 2]. For this we divide the interval [0, 2] in to four subintervals, namely $\Omega_1 = [0, 1 - \tau]$, $\Omega_2 = [1 - \tau, 1]$, $\Omega_3 = [1, 2 - \tau]$, $\Omega_4 = [2 - \tau, 2]$ where $\tau = \min\left\{0.5, \frac{2\varepsilon \ln(N)}{\alpha}\right\}$. Let $h = 2N^{-1}\tau$ and $H = 2N^{-1}(1 - \tau)$. The mesh $\overline{\Omega}^{2N} = \{x_0, x_1, \ldots, x_{2N}\}$ is defined by

$$\begin{cases} x_0 = 0.0, \\ x_i = x_0 + iH, \; i = 1(1)\frac{N}{2}, \qquad x_{i+\frac{N}{2}} = x_{\frac{N}{2}} + ih, \; i = 1(1)\frac{N}{2}, \\ x_{i+N} = x_N + iH, \; i = 1(1)\frac{N}{2}, \quad x_{i+\frac{3N}{2}} = x_{\frac{3N}{2}} + ih, \; i = 1(1)\frac{N}{2}. \end{cases}$$

9.1.3 A Hybrid Finite Difference Scheme for the TVP (16)

$$\begin{cases} L_1 V_i = 0, \quad i = 0(1)2N - 1 \\ V_{2N} = 1, \end{cases} \tag{19}$$

where

$$L_1 V_i = \begin{cases} \varepsilon \frac{V_{i+1} - V_i}{H} - a(x_i)V_i, & i = 0(1)\frac{N}{2} - 1, \\ \varepsilon \frac{V_{i+1} - V_i}{h} - a(x_i)V_i, & i = \frac{N}{2}(1)N - 1, \\ \varepsilon \frac{V_{i+1} - V_i}{h} - a(x_i)V_i, & i = N(1)\frac{3N}{2} - 1, \\ \varepsilon \frac{V_{i+1} - V_i}{h} - a(\frac{x_i + x_{i+1}}{2})\frac{V_i + V_{i+1}}{2}, & i = \frac{3N}{2}(1)2N - 1. \end{cases}$$

The following theorem gives an error estimate for this scheme.

Theorem 3 *Let $v_r(x)$ be the solution of (16). Further let V_i be its numerical solution defined by (19). Then*

$$| v_r(x_i) - V_i | \le CN^{-2} \ln^2 N, \quad i = 0(1)2N.$$

9.1.4 A Numerical Method for the Problem (15)

In order to obtain a numerical solution for the problem (15), we apply the fourth order Runge-Kutta method with piecewise cubic Hermite interpolation on $\overline{\Omega}^{2N}$. In fact we have

$$U_{0_{i+1}} = U_{0_i} + \frac{h^*}{6}(K_1 + 2K_2 + 2K_3 + K_4), \tag{20}$$

where,

$$h^* = \begin{cases} H, & i = 0(1)\frac{N}{2} - 1, \quad i = N(1)\frac{3N}{2} - 1, \\ h, & i = \frac{N}{2}(1)N - 1, \quad i = \frac{3N}{2}(1)2N - 1, \end{cases}$$

$$K_1 = \frac{h^*}{a(x_i)}\Big[f(x_i) - b(x_i)U_{0_i} - c(x_i)U_0^{h^*}(x_i)\Big],$$

$$K_2 = \frac{h^*}{a(x_i + \frac{h^*}{2})}\Big[f(x_i + \frac{h^*}{2}) - b(x_i + \frac{h^*}{2})(U_{0_i} + \frac{K_1}{2}) - c(x_i + \frac{h^*}{2})U_0^{h^*}(x_i + \frac{h^*}{2})\Big],$$

$$K_3 = \frac{h^*}{a(x_i + \frac{h^*}{2})}\Big[f(x_i + \frac{h^*}{2}) - b(x_i + \frac{h^*}{2})(U_{0_i} + \frac{K_2}{2}) - c(x_i + \frac{h^*}{2})U_0^{h^*}(x_i + \frac{h^*}{2})\Big],$$

$$K_4 = \frac{h^*}{a(x_i + h^*)}\Big[f(x_i + h^*) - b(x_i + h^*)(U_{0_i} + K_3) - c(x_i + h^*)U_0^{h^*}(x_i + h^*)\Big],$$

and

$$U_0^*(x) = \begin{cases} \phi(x-1), \; x \in [x_i, x_{i+1}], \; i = 0(1)N-1, \\ U_{0_{i-N}} A_{i-N}(x-1) + U_{0_{i-N+1}} A_{i+1-N}(x-1) + B_{i-N}(x-1) f^*(x_{i-N}) \\ \qquad + B_{i+1-N}(x-1) f^*(x_{i-N+1}), \quad x \in [x_i, x_{i+1}], \; i = N(1)2N-1, \end{cases}$$

$$A_i(x) = \left[1 - \frac{2(x - x_i)}{x_i - x_{i+1}}\right] \frac{(x - x_{i+1})^2}{(x_i - x_{i+1})^2},$$

$$A_{i+1}(x) = \left[1 - \frac{2(x - x_{i+1})}{x_{i+1} - x_i}\right] \frac{(x - x_i)^2}{(x_{i+1} - x_i)^2},$$

$$B_i(x) = \frac{(x - x_i)(x - x_{i+1})^2}{(x_i - x_{i+1})^2}, \quad B_{i+1}(x) = \frac{(x - x_{i+1})(x - x_i)^2}{(x_{i+1} - x_i)^2}$$

$$f^*(x_{i-N}) = \frac{f(x_{i-N})}{a(x_{i-N})} - \frac{b(x_{i-N})}{a(x_{i-N})}(U_{0_{i-N}}) - \frac{c(x_{i-N})}{a(x_{i-N})}\phi(x_{i-N} - 1),$$

The following theorem gives an **error estimate** for the above method.

Theorem 4 *Let $u_0(x)$ be the solution of the problem (15). Further let U_{0_i} be its numerical solution defined by (20). Then*

$$\mid u_0(x_i) - U_{0_i} \mid \leq C\overline{h}^4, \quad i = 0(1)2N,$$

where $\overline{h} = \max\ \{H, h\}$.

A numerical solution to the original problem (13) is given by

$$U_i = \begin{cases} U_{0_i} + x_i k_1, & i = 0(1)N, \\ U_{0_i} + V_i k_2 + k_3, & i = N+1(1)2N, \end{cases} \tag{21}$$

where U_{0_i} and V_i are numerical solutions of the problems (15) and (16) respectively and $k_1,\ k_2$ and k_3 are defined by (18). An error estimate for the above numerical solution is given in the following theorem.

Theorem 5 *Let $u(x)$ be the solution of the problem (13). Further let U_i be its numerical solution defined by (21). Then*

$$\mid u(x_i) - U_i \mid \leq C(\varepsilon + N^{-2}\ln^2 N), \quad i = 0(1)2N.$$

9.2 Asymptotic Numerical Method [21]

Asymptotic Numerical Method consists of 3 Steps:

- In the first step we obtain the reduced problem solution.
- In the second step we construct an auxiliary problem.
- A numerical solution is obtained in the third step.

Step 1: Solve the reduced problem of (13).

Theorem 6 *Let u be the solution of (13) and u_0 be its reduced problem solution as defined in (15). Then,*

$$\mid u(x) - u_0(x) \mid \leq C\varepsilon + C\exp(\frac{-\alpha(2-x)}{\varepsilon}), \quad x \in \overline{\Omega}.$$

Note: From the above theorem, it is clear that the solution u of the boundary value problem (13) exhibits a strong boundary layer at $x = 2$ and further, away from the boundary layer and in particular on [0, 1], we have

$$\mid u(x) - u_0(x) \mid \leq C\varepsilon + C\exp\left(\frac{-\alpha}{\varepsilon}\right), \quad x \in [0, 1].$$

Step 2: Define an auxiliary problem to (13):
Find $u^* \in Y$ such that

$$\begin{cases} P^*u^*(x): \ = -\varepsilon u^{*''}(x) + a(x)u^{*'}(x) + b(x)u^*(x) = f^*(x), \\ u^*(0) = u(0), \ u^*(2) = u(2), \end{cases} \tag{22}$$

where

$$f^*(x) = \begin{cases} f(x) - c(x)\phi(x-1), & x \in \Omega^- \cup \{1\}, \\ f(x) - c(x)u_0(x-1), & x \in \Omega^+. \end{cases}$$

We now state a maximum principle for this problem.

Theorem 7 *Let $w \in Y'$ be any function satisfying $w(0) \geq 0$, $w(2) \geq 0$, $P^*w(x) \geq 0$, $\forall x \in \Omega^- \cup \Omega^+$ and $[w'](1) \leq 0$. Then, $w(x) \geq 0, \quad \forall x \in \overline{\Omega}$.*

Theorem 8 *Let u and u^* be the solutions of the problems (13) and (22) respectively. Then,*

$$| u(x) - u^*(x) | \leq C\varepsilon + C \exp(\frac{-\alpha}{\varepsilon}), \quad x \in \overline{\Omega}.$$

Step 3:
On $\overline{\Omega}^{2N}$, we define the following scheme for the BVP (22):

$$\begin{cases} P^{*N}U^*(x_i) = -\varepsilon\delta^2 U^*(x_i) + a(x_i)D^- U^*(x_i) + b(x_i)U^*(x_i) = F^*(x_i), \\ \qquad\qquad\qquad\qquad\qquad\qquad\qquad i = 1(1)N-1, N+1(1)2N-1, \\ D^- U^*(x_N) = D^+ U^*(x_N), \\ U^*(x_0) = u^*(0), \ U^*(x_{2N}) = u^*(2), \end{cases} \tag{23}$$

where

$$\begin{aligned} \delta^2 U^*(x_i) &= \frac{2}{x_{i+1} - x_{i-1}} \left[D^+ U^*(x_i) - D^- U^*(x_i) \right], \\ D^- U^*(x_i) &= \frac{U^*(x_i) - U^*(x_{i-1})}{x_i - x_{i-1}}, \ D^+ U^*(x_i) = \frac{U^*(x_{i+1}) - U^*(x_i)}{x_{i+1} - x_i}, \\ F^*(x_i) &= f^*(x_i), \ x_i \in \overline{\Omega}^{2N} \setminus \{x_0, x_N, x_{2N}\}, \end{aligned} \tag{24}$$

or

$$F^*(x_i) = \begin{cases} f(x_i) - c(x_i)\phi(x_i - 1), \ x_i \in \Omega^- \cap \overline{\Omega}^{2N}, \\ f(x_i) - c(x_i)U_{0_{i-N}}, \ x_i \in \Omega^+ \cap \overline{\Omega}^{2N}. \end{cases} \tag{25}$$

Theorem 9 (Discrete Maximum Principle) *Suppose a mesh function $Z(x_i)$ satisfies $Z(x_0) \geq 0, Z(x_{2N}) \geq 0, P^{*N} Z(x_i) \geq 0, \ x_i \in \overline{\Omega}^{2N} \setminus \{x_0, x_N, x_{2N}\}$ and $[D]Z(x_N) = D^+ Z(x_N) - D^- Z(x_N) \leq 0$. Then, $Z(x_i) \geq 0, \ \forall x_i \in \overline{\Omega}^{2N}$.*

A consequence of this theorem is the following stability result.

Theorem 10 (Discrete Stability Result) *Let $U^*(x_i)$ be a numerical solution of the problem (23). Then,*

$$| U^*(x_i) | \leq C \max\{| U^*(x_0) |, | U^*(x_{2N}) |, \max_{j\in J} P^{*N} U^*(x_j)\},$$

$$J = \{1, \ldots, N-1, N+1, \ldots, 2N-1\},\ i = 0(1)2N.$$

Theorem 11 *Let u^* be the solution of the auxiliary problem (22) and let $U^*(x_i)$ be the corresponding numerical solution defined by (23) and (25). Then,*

$$| u^*(x_i) - U^*(x_i) | \leq CN^{-1} \ln N, \quad x_i \in \overline{\Omega}^{2N}.$$

Theorem 12 *Let $U^*(x_i)$ be a numerical solution of (22) defined by (23) and (25). Then,*

$$| u^*(x_i) - U^*(x_i) | \leq CN^{-1} \ln N, \quad x_i \in \overline{\Omega}^{2N}.$$

Theorem 13 *Let u be the solution of the problem (13) and let $U^*(x_i)$ be a numerical solution defined by (23) with either (25) or (25). Then,*

$$\| u - U^* \|_{\overline{\Omega}^{2N}} \leq CN^{-1} \ln N.$$

The above IVT and ANM can be applied to the following problems

- Convection diffusion equation with discontinuous source term
- Convection diffusion equation with discontinuous convection coefficient
- Neumann boundary value problem with smooth data
- Neumann boundary value problem with discontinuous source term
- System of convection diffusion equations with smooth data
- System of convection diffusion equations with discontinuous source terms
- Singularly perturbed third order delay differential equations [22]
- Singularly perturbed system of reaction-diffusion type delay differential equations [23]

9.3 *An Iterative Numerical Method [24]*

Consider BVP:
Find $u \in U := C^2(\Omega) \cap C(\overline{\Omega})$ such that

$$\begin{cases} -\varepsilon u''(x) + a(x)u(x) + b(x)u(x-1) = f(x), & x \in \Omega, \\ u(x) = \phi(x), \quad x \in [-1, 0], \\ u(2) = l, \end{cases} \tag{26}$$

where $0 < \alpha \leq a(x)$, $-\beta_0 \leq b(x) \leq \beta < 0$, for all $x \in \overline{\Omega}$, $\Omega = (0, 2)$, $\overline{\Omega} = [0, 2]$, $\alpha - \beta_0 > 0$, the functions a, b and f are sufficiently smooth on $\overline{\Omega}$ and ϕ is smooth on $[-1, 0]$. The above boundary value problem (26) has a solution and the solution is unique [25].

The above problem is equivalent to find $u \in U^* := C^2(\Omega^- \cup \Omega^+) \cap C(\overline{\Omega})$ such that

$$\begin{cases} Pu(x) := \begin{cases} -\varepsilon u''(x) + a(x)u(x) = f(x) - b(x)\phi(x-1), \ x \in \Omega^-, \\ -\varepsilon u''(x) + a(x)u(x) + b(x)u(x-1) = f(x), \ x \in \Omega^+, \end{cases} \\ u(0) = \phi(0), \ \ u(1^-) = u(1^+), \ \ u'(1^-) = u'(1^+), \ \ u(2) = l, \end{cases} \tag{27}$$

where $\Omega^- = (0, 1)$ and $\Omega^+ = (1, 2)$. This boundary value problem (27) exhibits strong boundary layers at $x = 0, \ x = 2$ and strong interior layers (left and right) at $x = 1$ [26].

9.3.1 Iterative Method

Following the method suggested in [25], we now suggest an iterative procedure for the boundary value problem (27) as follows. Let

$$\begin{cases} u_0(x) = \phi(x), & x \in [-1, 0], \\ u_0(x) = \phi(0), & x \in [0, 2], \end{cases} \tag{28}$$

and

$$\begin{cases} u_n(x) \in C^2(\Omega^- \cup \Omega^+) \cap C^1(\Omega) \cap C(\overline{\Omega}) \quad \text{such that} \\ u_n(x) = \phi(x), \quad x \in [-1, 0], \\ -\varepsilon u_n''(x) + a(x)u_n(x) = \begin{cases} f(x) - b(x)\phi(x-1), & x \in \Omega^-, \\ f(x) - b(x)u_{n-1}(x-1), & x \in \Omega^+, \end{cases} \\ u_n(0) = \phi(0), \quad u_n(2) = l. \quad \text{for } n = 1, \ 2, \ \cdots . \end{cases} \tag{29}$$

Theorem 14 *The sequence $\{u_n(x)\}$ defined by (28)–(29) converges uniformly to the solution u of the problem (27) on $\overline{\Omega}$.*

9.3.2 Shishkin Mesh

Since the boundary value problem (27) and the boundary value problems (29) exhibit boundary layers at $x = 0, \ x = 2$ and interior layers (left and right) at $x = 1$, we divide the interval $[0, 2]$ into six subintervals, namely $\Omega_1 = [0, \tau]$, $\Omega_2 = [\tau, 1 - \tau]$, $\Omega_3 = [1 - \tau, 1]$, $\Omega_4 = [1, 1 + \tau]$, $\Omega_5 = [1 + \tau, 2 - \tau]$ and $\Omega_6 = [2 - \tau, 2]$, where $\tau = \min\left\{\frac{1}{4}, \ \frac{2\sqrt{\varepsilon}\ln(N)}{\sqrt{\alpha}}\right\}$. Let $h = 4N^{-1}\tau$ and $H = 2N^{-1}(1 - 2\tau)$.

The Shishkin mesh $\overline{\Omega}^{2N} = \{x_0,\ x_1,\ \cdots,\ x_{2N}\}$ is defined by

$$\begin{cases} x_0 = 0.0,\ x_i = x_0 + ih,\ i = 1(1)\frac{N}{4},\ x_{i+\frac{N}{4}} = x_{\frac{N}{4}} + iH,\ i = 1(1)\frac{N}{2}, \\ x_{i+\frac{3N}{4}} = x_{\frac{3N}{4}} + ih,\ i = 1(1)\frac{N}{4},\ x_{i+N} = x_N + ih,\ i = 1(1)\frac{N}{4}, \\ x_{i+\frac{5N}{4}} = x_{\frac{5N}{4}} + iH,\ i = 1(1)\frac{N}{2},\ x_{i+\frac{7N}{4}} = x_{\frac{7N}{4}} + ih,\ i = 1(1)\frac{N}{4}. \end{cases}$$

9.3.3 Scheme

Using the finite difference scheme discussed in [27] on the Shishkin mesh $\overline{\Omega}^N = \{x_0,\ x_1,\ \cdots,\ x_N\}$, we now now define the following finite difference scheme for the sequence of the problems (29). Let $U^{[0]} = \big(u_0(x_0),\ u_0(x_1),\ \cdots,\ u_0(x_N)\big)$. Find $U^{[n]} = \big(U_0^{[n]},\ U_1^{[n]},\cdots,\ U_N^{[n]}\big)$ such that

$$\begin{cases} -\varepsilon\delta^2 U_i^{[n]} + a_i U_i^{[n]} = \begin{cases} f_i - b_i\phi(x_i - 1), & i = 1,\cdots,\frac{N}{2} - 1, \\ f_i - b_i U_{i-\frac{N}{2}}^{[n-1]}, & i = \frac{N}{2} + 1,\cdots, N-1, \end{cases} \\ U_0^{[n]} = \phi(0),\ U_N^{[n]} = l, \\ D^+ U_{\frac{N}{2}}^{[n]} = D^- U_{\frac{N}{2}}^{[n]}, \quad \text{for}\, n = 1,\ 2,\ \cdots. \end{cases} \tag{30}$$

Here

$$\delta^2 U_i^{[n]} = \frac{1}{x_{i+1} - x_{i-1}}\left(\frac{U_{i+1}^{[n]} - U_i^{[n]}}{x_{i+1} - x_i} - \frac{U_i^{[n]} - U_{i-1}^{[n]}}{x_i - x_{i-1}}\right)$$

$$D^+ U_{\frac{N}{2}}^{[n]} = \frac{U_{\frac{N}{2}+1}^{[n]} - U_{\frac{N}{2}}^{[n]}}{x_{\frac{N}{2}+1} - x_{\frac{N}{2}}} \quad \text{and} \quad D^- U_{\frac{N}{2}}^{[n]} = \frac{U_{\frac{N}{2}}^{[n]} - U_{\frac{N}{2}-1}^{[n]}}{x_{\frac{N}{2}} - x_{\frac{N}{2}-1}}.$$

9.3.4 Error Estimate

An error estimate for the above method is given as follows:

Theorem 15 *Let $u(x)$ and $U^{[n]}$ be the solutions of the problems (27) and (30) respectively. Then we have*

$$\|u - U^{[n]}\|_{\overline{\Omega}^N} \le CN^{-1}\ln N,$$

provided that

$$n \ge \frac{\ln(N^{-1}\ln N)}{\ln\gamma}, \quad \gamma = \frac{\beta_0}{\alpha}.$$

Following are the recommended papers for references. One may see some more references mentioned in these papers.

References

1. J.N. Kapur, *Mathematical Modelling* (New Age International Publishers, Delhi, 1988)
2. G.F. Simmons, *Differential Equations with Applications and Historical Notes* (Tata McGraw Hill, New Delhi, 2003)
3. R.V. Culshaw, Shigui Ruan, Glenn Webb, A mathematical model of cell-to-cell spread of HIV-1 that includes a time delay. J. Math. Biol. **46**, 425–444 (2003)
4. A.A. Samarskii, *The Theory of Difference Schemes* (Marcel Dekker Inc, 2001)
5. R. Bellman, K.L. Cooke, *Differential-Difference Equations* (Academic Press, 1963)
6. A. Bellen, M. Zennaro, *Numerical Methods for Delay Differential Equations* (Clarendon press, Oxford, 2003)
7. K.W. Chang, F.A. Howes, *Non-linear Singular Perturbation Phenomena: Theory and Application* (Springer, New York, 1984)
8. E.M. Dejager, J. Furu, *The Theory of Singular Perturbations* (Elsevier Science B.V, Netherlands, 1996)
9. A. Hasan Nayfeh, *Perturbation Methods* (Wiley, New York, 1973)
10. E.P. Doolan, J.J.H. Miller, W.H.A. Schilders, *Uniform Nerical Methods for Problems with Initial and Boundary Layers* (Boole, Dublin, 1980)
11. P. Singh, K.K. Sharma, Numerical solution of first-order hyperbolic partial differential-difference equation with shift. Num. Meth. Partial Diff. Eq. **26**(1), 107–116 (2010)
12. E. O'Riordan, G.I. Shishkin, Singularly perturbed parabolic problems with non-smooth data. J. Comput. Appl. Math. **166**, 233–245 (2004)
13. G.I. Shishkin, Grid approximation of singularly perturbed parabolic equations in the presence of weak and strong transient layers induced by a discontinuous right-hand side. Comput. Math. Math. Phys. **46**(3), 388–401 (2006)
14. Z. Cen, A Hybrid finite difference scheme for a class of singularly perturbed delay differential equations. Neural, Parallel, Scient. Comput. **16**, 303–308 (2008)
15. G.M. Amiraliyev, F. Erdogan, Uniform numerical method for singularly perturbed delay differential equations. Comput. Math. Appl. **53**, 1251–1259 (2007)
16. V. Subburayan, N. Ramanujam, An initial value technique for singularly perturbed reaction-diffusion problems with a negative shift. Novi Sad J. Math. **43**(2), 67–80 (2013)
17. V. Subburayan, N. Ramanujam, Asymptotic initial value technique for singularly perturbed convection diffusion delay problems with boundary and weak interior layers. Appl. Math. Lett. **25**, 2272–2278 (2012)
18. J.J.H. Miller, E.O. Riordan, G.I. Shishkin, *Fitted Numerical Method for Singular Perturbation Problems* (World Scientific, Singapore, 1996)
19. P.A. Farrell, A.F. Hegarty, J.J.H. Miller, E. O'Riordan, G.I. Shishkin, *Robust Computational Techniques for Boundary Layers* (Chapman Hall/CRC, Boca Raton, 2000)
20. V. Subburayan, N. Ramanujam, An initial value technique for singularly perturbed convection-diffusion problems with a delay. J. Opt. Theory Appl **158**, 234–250 (2013)
21. V. Subburayan, N. Ramanujam, An asymptotic numerical method for singularly perturbed convection-diffusion problems with a negative shift. Neural, Parallel Sci. Comput. **21**, 431–446 (2013)
22. V. Subburayan, N. Ramanujam, *An Initial value method for singularly perturbed third order delay differential equations*, International Conference on Mathematical Sciences (ICMS), July 17-1-9, 2014, Sathyabhama University, Chennai, India, pp. 221–229, ISBN:978-93-5107-261-4, Elsevier
23. V. Subburayan, N. Ramanujam, An initial value method for singularly perturbed system of reaction-diffusion type delay differential equations. J. KSIAM. **17**(4), 221–237 (2013)
24. P. Avudai Selvi, N. Ramanujam, *An Iterative Numerical Method for Singularly Perturbed Reaction-diffusion Equations with Negative Shift* (communicated)
25. P. Chocholaty, L. Slahor, A numerical method to boundary value problems for second order delay differential equations. Numer. Math. **33**, 69–75 (1979)

26. C.G. Lange, R.M. Miura, Singularly perturbation analysis of boundary-value problems for differential-difference equations. SIAM. J. Appl. Math. **42**(3), 502–531 (1982)
27. P.A. Farrell, J.J.H. Miller, E. O'Riordan, G.I. Shishkin, *Singularly perturbed differential equations with discontinuous source terms*. In G.I.Shishkin, J.J.H. Miller, L. Vulkov (eds.) Analytical and numerical methods for convection-dominated and singularly perturbed problems (Nova Science Publishers. Inc, New York, USA, 2000), pp. 23-32

Initial or Boundary Value Problems for Systems of Singularly Perturbed Differential Equations and Their Solution Profile

Valarmathi Sigamani

Abstract Singular perturbation problems, by nature, are not easy to handle and they demand efficient techniques to solve and careful analysis. And systems of singular perturbation problems are tougher as their solutions exhibit layers with sub-layers. Their properties are studied and examples are given to illustrate.

Keywords Singular perturbation problems · Initial/boundary layers · Sublayers · Shishkin meshes · Finite difference scheme · Parameter uniform convergence

1 Introduction

Recently systems of singularly perturbed differential equations are studied by many researchers all over the world. To cite a few: [1–20]. Most of the works are confined to systems with two equations and a few works are found on systems of n equations; $n > 0$ is arbitrary. Here, three types of systems of singularly perturbed differential equations are to be discussed.

2 A System of First Order Ordinary Differential Equations

Consider the system

$$E\mathbf{u}'(x) + A(x)\mathbf{u}(x) = \mathbf{f}(x), \quad x \in \Omega = (0, X] \tag{1}$$

with $\mathbf{u}(0) = \boldsymbol{\phi}$ given. E is the diagonal matrix $E = diag(\varepsilon_i), i = 1, 2, \cdots, n$ and $A(x) = (a_{ij}(x))$ is an $n \times n$ matrix. The functions $a_{ij}(x)$ and $f_i(x)$ for $1 \leq i, j \leq n$ are assumed to be in $C^2(\overline{\Omega})$ where $\overline{\Omega} = [0, 1]$, assuming, without loss of

V. Sigamani (✉)
Department of Mathematics, Bishop Heber College, Tiruchirappalli, Tamil Nadu, India
e-mail: valarmathi07@gmail.com

V. Sigamani et al. (eds.), *Differential Equations and Numerical Analysis*,
Springer Proceedings in Mathematics & Statistics 172,
DOI 10.1007/978-81-322-3598-9_4

generality, $X = 1$. For convenience, the ordering $\varepsilon_1 < \varepsilon_2 < \cdots \varepsilon_n$ is assumed. Further, the functions a_{ij} are assumed to satisfy

$$a_{ii}(x) > \sum_{\substack{j=1 \\ j \neq i}}^{n} |a_{ij}(x)|, i = 1, 2, \cdots, n \tag{2}$$

$$a_{ij}(x) \leq 0, 1 \leq i \neq j \leq n \tag{3}$$

and the singular perturbation parameters $\varepsilon_i, i = 1, 2, \cdots, n$ are assumed to satisfy

$$\varepsilon_n \leq \frac{\alpha}{6} \tag{4}$$

so as to accommodate all the layers well inside the domain.

With the above assumptions, the problem (1) has a solution $\mathbf{u} \in C^{(0)}(\overline{\Omega}) \cap C^{(1)}(\Omega)$

As explained in [21], here also the supremum norm is used in estimates. The norms $||\mathbf{V}|| = \max_{1 \leq k \leq n} |V_k|$ for any $n-$vector $\mathbf{V}$, $||y|| = \sup_{0 \leq x \leq 1} |y(x)|$ for any scalar-valued function y and $||\mathbf{y}|| = \max_{1 \leq k \leq n} ||y_k||$ for any vector valued function $\mathbf{y}$ are introduced.

The problem (1) is singularly perturbed, in the following sense. The reduced problem obtained by putting each $\varepsilon_i = 0,\ i = 1, 2, \cdots, n$, in (1) is the linear algebraic system

$$A(x)\mathbf{u}_0(x) = \mathbf{f}(x). \tag{5}$$

This problem (5) has a unique solution and hence arbitrary initial conditions cannot be imposed. This shows that there are initial layers at $x = 0$ for $\mathbf{u}$. The attracting feature of the layers is that the component u_n has an initial layer of width $O(\varepsilon_n)$, the component u_{n-1} also has a layer of width $O(\varepsilon_n)$ and an additional sublayer of width $O(\varepsilon_{n-1})$ and so on. Lastly the component u_1 has an initial layer of width $O(\varepsilon_n)$ and additional sub-layers of widths $O(\varepsilon_{n-1}), O(\varepsilon_{n-2}), \cdots, O(\varepsilon_1)$. The complexity of the layer pattern of the solution makes the problem more interesting. This complexity makes the derivation of bounds on the estimates of the derivatives and the error analysis more challenging.

2.1 Analytical Results

Valarmathi and Miller [19] established the maximum principle for a general system of $'n'$ linear first order singularly perturbed differential equations, with an additional result that the maximum principle satisfied by the operator

$\mathbf{L} = ED + A(x)$ of system (1) implies that the operator $\tilde{\mathbf{L}}$ of any lower order system also satisfies the maximum principle.

Apart from the stability result, estimates of the derivatives of smooth and singular components derived with the help of induction will not suffice for error analysis. Novel estimates of derivatives are required. To achieve this, points of interaction of the layer functions are identified. For a system of two equations, it was Linss [9] who identified such a point. But it is Valarmathi and Miller [20] who identified a sequence of such points between the 'n' layer functions and came out with some interesting properties, which lead to non classical bounds for the derivatives of the singular components that are interlinked.

2.2 Shishkin Mesh

The construction of an appropriate mesh plays a vital role in solving the singular perturbation problem. As there are layer regions or inner regions and outer regions and as more information is needed inside the inner region, a piecewise uniform mesh is needed.

A piecewise uniform Shishkin mesh distributing $N/2$ points to the outer region and the remaining $N/2$ points equally to all the inner regions will serve the purpose. The Shishkin mesh suggested for problem (1) is the set of points $\overline{\Omega}^N = \{x_j\}_0^N$ that divides $[0, 1]$ into $n+1$ mesh intervals $[0, \sigma_1] \cup ... \cup (\sigma_{n-1}, \sigma_n] \cup (\sigma_n, 1]$ where the n parameters σ_r separate the uniform meshes. With $\sigma_0 = 0, \sigma_{n+1} = 1, \sigma_n$ is defined by $\sigma_n = \min\left\{\frac{\sigma_{n+1}}{2}, \frac{\varepsilon_n}{\alpha}\ln N\right\}$ and for $r = n-1, n-2, ..., 2, 1,\ \sigma_r = \min\left\{\frac{r\sigma_{r+1}}{r+1}, \frac{\varepsilon_r}{\alpha}\ln N\right\}$. Then on the subinterval $(\sigma_n, 1]$, a uniform mesh with $N/2$ mesh points is placed and on each of the intervals $(\sigma_r, \sigma_{r+1}], r = 0, 1, ..., n-1$, a uniform mesh of $N/2$n mesh points is placed where $'n'$ is the number of perturbation parameters involved in (1).

In particular, when all the parameters $\sigma_r, r = 1, 2, ..., n$ are with the left choice, the Shishkin mesh becomes a classical uniform mesh with stepsize N^{-1} through out from 0 to 1. For the other cases, the mesh is coarse in the outer region and becomes finer and finer towards the initial point. Infact $\sigma_r, r = 1, 2, 3, \cdots, n$ are the points only where a change in the mesh size may occur.

2.3 Discrete Problem

To solve (1) numerically, consider the corresponding discrete initial value problem on the Shishkin mesh $\overline{\Omega}^N$ given by

$$ED^{-}\mathbf{U} + A\mathbf{U} = \mathbf{f} \text{ on } \Omega^N, \mathbf{U} = \mathbf{u} \text{ at the initial point.} \tag{6}$$

Making use of the mesh geometry and the novel estimates of derivatives derived by the existence of the sequence of layer interaction points, the authors in [20] established the almost first order parameter uniform convergence.

More general case of problem (1)

In nature, many systems of multiscale dynamics, involve some components having large scale flow rates. This problem when formulated follows the form $E D\mathbf{u} + A\mathbf{u} = \mathbf{f}$ on $(0, 1]$ and $\mathbf{u}(0) = \boldsymbol{\phi}$ where $E =$ diag(ε_i) with $0 < \varepsilon_1 < \varepsilon_2 < \ldots < \varepsilon_k = \varepsilon_{k+1} = \ldots = \varepsilon_n = 1$. In this case, the problem is called a partially singularly perturbed initial value problem for a linear system of first order ODEs.

Establishing analytical results and error analysis demand the judicial use of certain barrier functions and the appropriate modification of the Shishkin mesh considered for problem (1). In the construction of the Shishkin mesh for solving problem (1), the number of transition parameters was fixed to be equal to the number of distinct perturbation parameters in (1). Here also, having the same strategy, the outer region gets wider as the number of transition parameters gets reduced.

2.4 Discontinuous Source Terms

In some multiscale fluid flows, it may also happen that some of the source functions f_i, $1 \le i \le n$ may go discontinuous at points in the domain of definition of problem (1). These discontinuities result in some interesting characteristics of the solution.

The solution, apart from its initial layers, exhibits interior layers at the points of discontinuity. Then care has to be taken in constructing the mesh because it should resolve interior layers in addition to the initial layers. Further, for a simple discontinuity at a point, the interior layers are just like the initial layers dislocated. These layer functions have a similar sequence of layer interaction points. Making use of these facts and the mesh geometry one can solve the problem with discontinuous source function.

Example 1 Consider the following system of singularly perturbed initial value problem.

$$\left.\begin{aligned} \varepsilon_1 u_1'(t) + 2(1+t)^2 u_1(t) - (1+t^2) u_2(t) &= 0.5(1+t) \\ \varepsilon_2 u_2'(t) - (1+t) u_1(t) + 2(1+t) u_2(t) &= (1+\frac{t}{4}) \end{aligned}\right\}$$

for $t \in (0, 1]$ and $\mathbf{u}(0) = \mathbf{0}$. The layer profile of the solution $\mathbf{u}$ of this problem obtained by the proposed method is as in Fig. 1 for $\varepsilon_1 = 10^{-10}$, $\varepsilon_2 = 10^{-7}$ and $N = 128$.

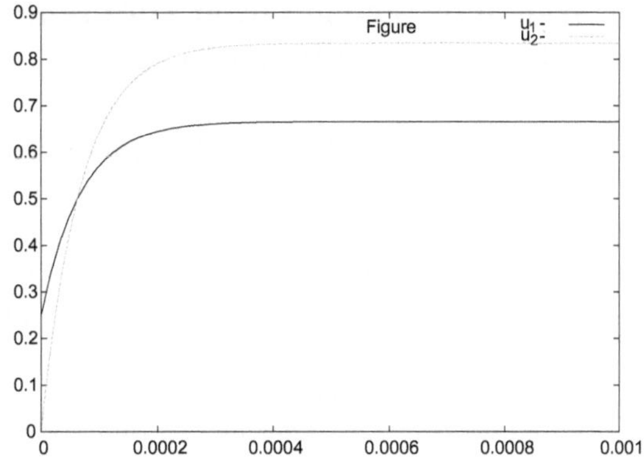

Fig. 1 Solution profile of Example 1

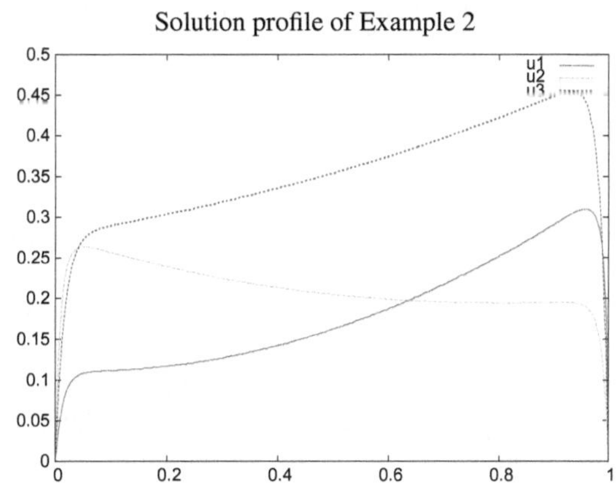

Fig. 2 Solution profile of Example 2

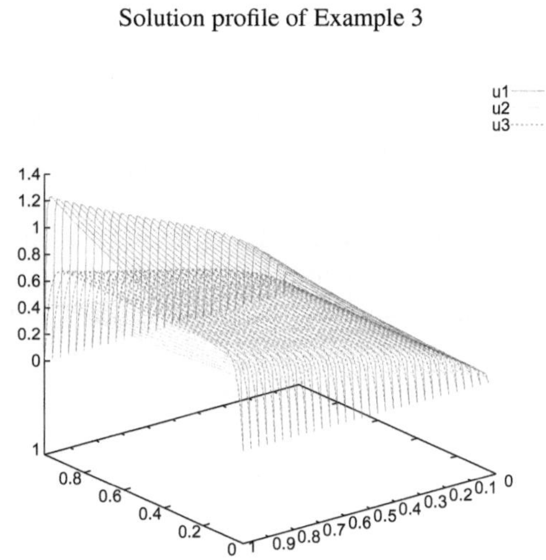

Fig. 3 Solution profile of Example 3

3 System of Second Order Differential Equations of Reaction—Diffusion Type

Consider the system of singularly perturbed differential equations of reaction-diffusion type with boundary values prescribed.

$$-E\mathbf{u}''(x) + A(x)\mathbf{u}(x) = \mathbf{f}(x),\ x \in \Omega = (0, 1)$$
$$\mathbf{u}(0), \mathbf{u}(1) \text{ given} \tag{7}$$

E is the same as in problem (1), $A = (a_{ij})_{n\times n},\ a_{ij}(x),\ f_i(x) \in C^2(\overline{\Omega})$ and (2), (3) & (4) hold good in $\overline{\Omega}$. Under these assumptions the problem (7) has a solution in $C^{(0)}(\overline{\Omega}) \cap C^{(2)}(\Omega)$.

For systems of this type, Paramasivam et al. [16] established maximum principle, the analytical results and a parameter uniform method of solving them on a Shishkin mesh.

The solution $\mathbf{u}$ of the problem (7) exhibits twin boundary layers at the boundary, $x = 0$ and $x = 1$. The component u_n exhibits twin boundary layers of width $O(\sqrt{\varepsilon_n})$, while u_{n-1} has twin boundary layers of width $O(\sqrt{\varepsilon_n})$ and additional twin boundary sub layers of width $O(\sqrt{\varepsilon_{n-1}})$ and so on. Lastly u_1 has twin boundary layers of width $O(\sqrt{\varepsilon_n})$ and additional twin boundary sub layers of widths $O(\sqrt{\varepsilon_{n-1}}), O(\sqrt{\varepsilon_{n-2}}), \cdots, O(\sqrt{\varepsilon_1})$.

These boundary layers also have twin layer interaction sequences which could be used with the induction method in establishing the novel estimates of the derivatives of the smooth and singular components of the solution.
The related systems of (7) which are partially singularly perturbed and which have discontinuous source vector are with higher order difficulty and are handled as in the previous case, in [22, 23].

Example 2 Consider the following singularly perturbed boundary value problem

$$\left.\begin{array}{lcl} -\varepsilon_1 u_1''(x) + 5u_1(x) - u_2(x) - u_3(x) & = & x^2 \\ -\varepsilon_2 u_2''(x) - u_1(x) + (5+x)u_2(x) - u_3(x) & = & e^{-x} \\ -\varepsilon_3 u_3''(x) - (1+x)u_1(x) - u_2(x) + (5+x)u_3(x) & = & 1+x \end{array}\right\}$$

for $x \in (0, 1)$ and $\mathbf{u}(0) = \mathbf{0},\ \mathbf{u}(1) = \mathbf{0}$. The layer profile of the solution $\mathbf{u}$ of this problem obtained by the method suggested in [16] is presented in Fig. 2 for $\varepsilon_1 = \dfrac{\eta}{16}, \varepsilon_2 = \dfrac{\eta}{4}, \varepsilon_3 = \eta$ where $\eta = 0.1$ and $N = 512$.

4 Systems of Singularly Perturbed Time Dependent Equations of Reaction-Diffusion Type

Consider the following parabolic initial-boundary value problem for a singularly perturbed linear system of second order differential equations

$$\frac{\partial \mathbf{u}}{\partial t} - E\frac{\partial^2 \mathbf{u}}{\partial x^2} + A\mathbf{u} = \mathbf{f}, \text{ on } \Omega, \ \mathbf{u} \text{ given on } \Gamma, \tag{8}$$

where $\Omega = \{(x,t) : 0 < x < 1, 0 < t \le T\}$, $\overline{\Omega} = \Omega \cup \Gamma$, $\Gamma = \Gamma_L \cup \Gamma_B \cup \Gamma_R$ with $\mathbf{u}(0,t) = \boldsymbol{\phi}_L(t)$ on $\Gamma_L = \{(0,t) : 0 \le t \le T\}$, $\mathbf{u}(x,0) = \boldsymbol{\phi}_B(x)$ on $\Gamma_B = \{(x, 0) : 0 < x < 1\}$, $\mathbf{u}(1,t) = \boldsymbol{\phi}_R(t)$ on $\Gamma_R = \{(1,t) : 0 \le t \le T\}$. Here, for all $(x,t) \in \overline{\Omega}$, $\mathbf{u}(x,t) = (u_1(x,t), u_2(x,t), \dots, u_n(x,t))^T$, $\mathbf{f}(x,t) = (f_1(x,t), f_2(x,t), \dots, f_n(x,t))^T$,

$$E = \begin{pmatrix} \varepsilon_1 & 0 & \cdots & 0 \\ 0 & \varepsilon_2 & \cdots & 0 \\ \vdots & \vdots & & \vdots \\ 0 & 0 & \cdots & \varepsilon_n \end{pmatrix}, \quad A(x,t) = \begin{pmatrix} a_{11}(x,t) & a_{12}(x,t) & \cdots & a_{1n}(x,t) \\ a_{21}(x,t) & a_{22}(x,t) & \cdots & a_{2n}(x,t) \\ \vdots & \vdots & & \vdots \\ a_{n1}(x,t) & a_{n2}(x,t) & \cdots & a_{nn}(x,t) \end{pmatrix}.$$

The problem (8) can also be written in the operator form

$$\mathbf{L}\mathbf{u} = \mathbf{f} \text{ on } \Omega, \ \mathbf{u} \text{ given on } \Gamma,$$

where the operator $\mathbf{L}$ is defined by

$$\mathbf{L} = I\frac{\partial}{\partial t} - E\frac{\partial^2}{\partial x^2} + A,$$

where I is the identity matrix.

The reduced problem obtained by putting $\varepsilon_i = 0, \ i = 1, 2, \cdots, n$ in (8) is defined by

$$\frac{\partial \mathbf{u}_0}{\partial t} + A\mathbf{u}_0 = \mathbf{f}, \text{ on } \Omega, \ \mathbf{u}_0 = \mathbf{u} \text{ on } \Gamma_B. \tag{9}$$

The ε_i are assumed to be distinct and, for convenience, to have the ordering $\varepsilon_1 < \cdots < \varepsilon_n$. For all $(x,t) \in \overline{\Omega}$, it is assumed that the components $a_{ij}(x,t)$ of $A(x,t)$ satisfy the inequalities

$$a_{ii}(x,t) > \sum_{\substack{j \ne i \\ j=1}}^{n} |a_{ij}(x,t)| \text{ for } 1 \le i \le n, \text{ and } a_{ij}(x,t) \le 0 \text{ for } i \ne j \tag{10}$$

and there exists a number α satisfying the inequality $0 < \alpha < \min_{\substack{(x,t)\in\overline{\Omega} \\ 1\le i\le n}} (\sum_{j=1}^{n} a_{ij}(x,t))$.

It is also assumed, without loss of generality, that $\sqrt{\varepsilon_n} \le \frac{\sqrt{\alpha}}{6}$ which ensures that the solution domain contains all the layers.

The norms, $\parallel \mathbf{V} \parallel = \max_{1\le k\le n} |V_k|$ for any n-vector $\mathbf{V}$, $\parallel y \parallel_D = \sup\{|y(x,t)| : (x,t) \in D\}$ for any scalar-valued function y and domain D, and $\parallel \mathbf{y} \parallel = \max_{1\le k\le n} \parallel y_k \parallel$ for any vector-valued function $\mathbf{y}$, are introduced. When $D = \overline{\Omega}$ or Ω the subscript D is usually dropped. In a compact domain D a function is said to be Hölder continuous of degree λ, $0 < \lambda \le 1$, if, for all $(x_1,t_1), (x_2,t_2) \in D$,

$$|u(x_1,t_1) - u(x_2,t_2)| \le C(|x_1 - x_2|^2 + |t_1 - t_2|)^{\lambda/2}.$$

The set of Hölder continuous functions forms a normed linear space $C^0_\lambda(D)$ with the norm

$$||u||_{\lambda,D} = ||u||_D + \sup_{(x_1,t_1),(x_2,t_2)\in D} \frac{|u(x_1,t_1) - u(x_2,t_2)|}{(|x_1 - x_2|^2 + |t_1 - t_2|)^{\lambda/2}},$$

where $||u||_D = \sup_{(x,t)\in D} |u(x,t)|$. For each integer $k \ge 1$, the subspaces $C^k_\lambda(D)$ of $C^0_\lambda(D)$, which contain functions having Hölder continuous derivatives, are defined as follows

$$C^k_\lambda(D) = \{u : \frac{\partial^{l+m} u}{\partial x^l \partial t^m} \in C^0_\lambda(D) \text{ for } l, m \ge 0 \text{ and } 0 \le l + 2m \le k\}.$$

The norm on $C^0_\lambda(D)$ is taken to be $||u||_{\lambda,k,D} = \max_{0\le l+2m\le k} ||\frac{\partial^{l+m} u}{\partial x^l \partial t^m}||_{\lambda,D}$. For a vector function $\mathbf{v} = (v_1, v_2, ..., v_n)$, the norm is defined by $||\mathbf{v}||_{\lambda,k,D} = \max_{1\le i\le n} ||v_i||_{\lambda,k,D}$.

Regularity and Compatibility conditions

It is assumed that enough regularity and compatibility conditions hold for the data of the problem (8) so that the partial derivatives with respect to the space variable of the solution are continuous up to fourth order and the partial derivatives with respect to the time variable of the solution are continuous up to second order. The compatibility conditions for the problem (8) defined on a rectangular domain Ω is established in [3].

Sufficient conditions for the existence, uniqueness and regularity of solution of (8) are given in the following.

Assume that A, $\mathbf{f} \in C^2_\lambda(\overline{\Omega})$, $\boldsymbol{\phi}_L \in C^1(\Gamma_L)$, $\boldsymbol{\phi}_B \in C^2(\Gamma_B)$, $\boldsymbol{\phi}_R \in C^1(\Gamma_R)$ and that the following compatibility conditions are fulfilled at the corners $(0,0)$ and $(1,0)$ of Γ

$$\boldsymbol{\phi}_B(0) = \boldsymbol{\phi}_L(0) \quad \text{and} \quad \boldsymbol{\phi}_B(1) = \boldsymbol{\phi}_R(0), \tag{11}$$

$$\begin{aligned}\frac{d\boldsymbol{\phi}_L(0)}{dt} - E\frac{d^2\boldsymbol{\phi}_B(0)}{dx^2} + A(0,0)\boldsymbol{\phi}_B(0) = \mathbf{f}(0,0),\\ \frac{d\boldsymbol{\phi}_R(0)}{dt} - E\frac{d^2\boldsymbol{\phi}_B(1)}{dx^2} + A(1,0)\boldsymbol{\phi}_B(1) = \mathbf{f}(1,0)\end{aligned} \tag{12}$$

and

$$\begin{aligned}\frac{d^2}{dt^2}\boldsymbol{\phi}_L(0) = E^2\frac{d^4}{dx^4}\boldsymbol{\phi}_B(0) - 2EA(0,0)\frac{d^2}{dx^2}\boldsymbol{\phi}_B(0) - EA(0,0)\frac{d}{dx}\boldsymbol{\phi}_B(0)\\ -(A^2(0,0) + \frac{\partial A}{\partial t}(0,0) + E\frac{\partial^2 A}{\partial x^2}(0,0))\boldsymbol{\phi}_B(0)\\ -A(0,0)\mathbf{f}(0,0) + \frac{\partial \mathbf{f}}{\partial t}(0,0) + E\frac{\partial^2 \mathbf{f}}{\partial x^2}(0,0),\end{aligned} \tag{13}$$

$$\begin{aligned}\frac{d^2}{dt^2}\boldsymbol{\phi}_R(0) = E^2\frac{d^4}{dx^4}\boldsymbol{\phi}_B(1) - 2EA(1,0)\frac{d^2}{dx^2}\boldsymbol{\phi}_B(1) - EA(1,0)\frac{d}{dx}\boldsymbol{\phi}_B(1)\\ -(A^2(1,0) + \frac{\partial A}{\partial t}(1,0) + E\frac{\partial^2 A}{\partial x^2}(1,0))\boldsymbol{\phi}_B(1)\\ -A(1,0)\mathbf{f}(1,0) + \frac{\partial \mathbf{f}}{\partial t}(1,0) + E\frac{\partial^2 \mathbf{f}}{\partial x^2}(1,0).\end{aligned} \quad ss \tag{14}$$

Then there exists a unique solution $\mathbf{u}$ of (8) satisfying $\mathbf{u} \in C_\lambda^4(\overline{\Omega})$.

As there are twin boundary parabolic layers with sub-layers, the Shishkin mesh to resolve these layers is constructed on the rectangular domain $\overline{\Omega}$ and a classical finite difference method is suggested and proved to be parameter-uniform first order convergent in time and almost second order convergent in space in [3].

Example 3 Consider the problem

$$\frac{\partial \mathbf{u}}{\partial t} - E\frac{\partial^2 \mathbf{u}}{\partial x^2} + A\mathbf{u} = \mathbf{f} \text{ on } (0,1)\times(0,1], \quad \mathbf{u} = \mathbf{0} \text{ for } x = 0 \text{ or } t = 0 \text{ or } x = 1,$$

where $E = diag(\varepsilon_1, \varepsilon_2, \varepsilon_3)$, $A = \begin{pmatrix} 6 & -1 & 0 \\ -t & 5(x+1) & -1 \\ -1 & -(1+x^2) & 6+x \end{pmatrix}$, $\mathbf{f} = \begin{pmatrix} 1+e^{x+t} \\ 1+x+t^2 \\ 1+e^t \end{pmatrix}$.
The layer profile of the solution $\mathbf{u}$ of this problem is displayed in Fig. 3 for $\varepsilon_1 = 2^{-7}, \varepsilon_2 = 2^{-5}, \varepsilon_3 = 2^{-2}, M = 32$ and $N = 48$.

Here for the system (8) also, its subcases of the system being partially perturbed and the source vector to have discontinuities could also be dealt with in a way similar to those in Sects. 2 and 3.

References

1. C. Clavero, J.L. Gracia, F.J. Lisbona, An almost third order finite difference scheme for singularly perturbed reaction-diffusion systems. J. Comput. Appl. Math. **234**, 25012515 (2010)
2. C. Clavero, F.J. Lisbona J.L. Gracia, Second order uniform approximations for the solution of time dependent singularly perturbed reaction diffusion systems. Int. J. Numer. Anal. Mod. **7**, 428443 (2010)
3. V. Franklin, M. Paramasivam, J.J.H. Miller, S. Valarmathi, Second order parameter-uniform convergence for a finite difference method for a singularly perturbed linear parabolic system. Int. J. Numer. Anal. Model. **10**(1), 178–202
4. V. Franklin, J.J.H. Miller, S. Valarmathi, Second order parameter-uniform convergence for a finite difference method for a partially singularly perturbed linear parabolic system. Math. Commun. **19**, 469–496 (2014)
5. J.L. Gracia, F.J. Lisbona, A uniformly convergent scheme for a system of reaction-diffusion equations. J. Comput. Appl. Math. **206**(1), 116 (2007)
6. J.L. Gracia, F.J. Lisbona, M. Madaune-Tort, E. ORiordan, A system of singularly perturbed semilinear equations. Lect. Notes Comput.Sci. Eng.Springer, **69**, 163172 (2009)
7. J.L. Gracia, F.J. Lisbona, E. Oriordan, A coupled system of singularly perturbed parabolic reaction-diffusion equations. Adv. Comput. Math. **32**(1), 4361 (2010)
8. T. Linβ, A robust method for a time-dependent system of coupled singularly perturbed reaction-diffusion problems. Technical report, MATHNM- 03-2008, TU Dresdon, 2008
9. T. Linss, N. Madden, Accurate solution of a system of coupled singularly perturbed reaction-diffusion equations. Computing **73**, 121133 (2004)
10. T. Linβ, N. Madden, Parameter uniform approximations for timedependent reaction-diffusion problems. Numer. Meth. Partial Differ. Eq. **23**, 12901300 (2007)
11. T. Linss, N. Madden, Layer-adapted meshes for a linear system of coupled singularly perturbed reaction-diffusion problems. IMA J. Numer. Anal. **29**(1), 109125 (2009)
12. T. Linβ, N. Madden, Analysis of an alternating direction method applied to singularly perturbed reaction-diffusion problems. Int. J. Numer. Anal. Model. **7**(3), 507519 (2010)
13. T. Linss, M. Stynes, Numerical solution of systems of singularly perturbed differential equations. Comput. Meth. Appl. Math. **9**(2), 165191 (2009)
14. S. Matthews, E. ORiordan, G.I. Shishkin, A numerical method for a system of singularly perturbed reaction-diffusion equations. J. Comp. Appl. Math. **145**, 151166 (2002)
15. N. Madden, M. Stynes, A uniformly convergent numerical method for a coupled system of two singularly perturbed linear reaction-diffusion problems. IMA J. Numer. Anal. **23**(4), 627644 (2003)
16. M. Paramasivam, S. Valarmathi, J.H. John, Miller. Second order parameter uniform convergence for a finite difference method for a singularly perturbed linear reaction diffusion system. Math. Commun. **15**, 587612 (2010)
17. G.I. Shishkin, Approximation of systems of singularly perturbed elliptic reaction-diffusion equations with two parameters. Comput. Math. Math. Phys. **47**(5), 797828 (2007)
18. M. Stephens, N. Madden, A parameter-uniform schwarz method for a coupled system of reaction-diffusion equations. Journal of Computational and Applied Mathematics **230**(2), 360370 (2009)
19. S. Valarmathi, J.J.H. Miller, A parameter-uniform finite difference method for a singularly perturbed initial value problem: a special case. Lect. Notes Comput. Sci. Eng. Springer-Verlag **29**, 267276 (2009)
20. S. Valarmathi, J.J.H. Miller, A parameter-uniform finite difference method for singularly perturbed linear dynamical systems. Int. J. Numer. Anal. Model. **7**(3), 535548 (2010)
21. J.J.H. Miller, E.O Riordan, G.I. Shishkin, *Fitted Numerical Methods for Singular Perturbation Problems*. Error estimates in the maximum norm for linear problems in one and two dimensions (World Scientific publishing CO.Pvt.Ltd., Singapore, 1996)

22. M. Paramasivam, S. Valarmathi, J.J.H. Miller, Parameter-uniform convergence for a finite difference method for a singularly perturbed linear reaction-diffusion system with discontinuous source term. Math. Commun. **15**(2), 587612 (2010)
23. M. Paramasivam, J.J.H. Miller, S. Valarmathi, Second order parameter-uniform numerical method for a partially singularly perturbed linear system of reaction-diffusion type. Math. Commun. **18**, 271295 (2013)

Part II
Contributed Papers

Convergence of the Crank-Nicolson Method for a Singularly Perturbed Parabolic Reaction-Diffusion System

Franklin Victor, John J.H. Miller and Valarmathi Sigamani

Abstract A general parabolic system of singularly perturbed linear equations of reaction-diffusion type is considered. The components of the solution exhibit overlapping layers. A numerical method with the Crank-Nicolson operator on a uniform mesh for time and classical finite difference operator on a Shishkin piecewise uniform mesh for space is constructed. It is proved that in the maximum norm, the numerical approximations obtained with this method are second order convergent in time and essentially second order convergent in space.

Keywords Singular perturbation problems · Parabolic problems · Boundary layers · Uniform convergence · Finite difference scheme · Shishkin mesh

1 Introduction

The following parabolic initial-boundary value problem is considered

$$\mathbf{Lu} = \frac{\partial \mathbf{u}}{\partial t} + \mathbf{L}_x\mathbf{u} = \mathbf{f} \text{ on } \Omega, \quad \mathbf{u} \text{ given on } \Gamma, \tag{1}$$

where the operator $\mathbf{L}_x$ is defined by

$$\mathbf{L}_x = -E\frac{\partial^2}{\partial x^2} + A.$$

F. Victor (✉) · V. Sigamani
Department of Mathematics, Bishop Heber College, Tiruchirappalli, Tamil Nadu, India
e-mail: franklinvicto@gmail.com

V. Sigamani
e-mail: valarmathi07@gmail.com

J.J.H. Miller
Institute for Numerical Computation and Analysis, 7-9 Dame Court, Dublin 2, Ireland
e-mail: jm@incaireland.org

V. Sigamani et al. (eds.), *Differential Equations and Numerical Analysis*,
Springer Proceedings in Mathematics & Statistics 172,
DOI 10.1007/978-81-322-3598-9_5

Here $\Omega = \{(x,t) : 0 < x < 1, 0 < t \le T\}$, $\overline{\Omega} = \Omega \cup \Gamma$, $\Gamma = \Gamma_L \cup \Gamma_B \cup \Gamma_R$ where $\Gamma_L = \{(0,t) : 0 \le t \le T\}$, $\Gamma_B = \{(x,0) : 0 \le x \le 1\}$ and $\Gamma_R = \{(1,t) : 0 \le t \le T\}$. For all $(x,t) \in \overline{\Omega}$, $\mathbf{u}(x,t)$ and $\mathbf{f}(x,t)$ are column $n-$ vectors, E and $A(x,t)$ are $n \times n$ matrices, $E = \text{diag}(\boldsymbol{\varepsilon})$, $\boldsymbol{\varepsilon} = (\varepsilon_1, \cdots, \varepsilon_n)$ with $0 < \varepsilon_i < 1$ for all $i = 1, \ldots, n$. The ε_i are assumed, for convenience, to have the ordering $\varepsilon_1 \le \cdots \le \varepsilon_n$. For all $(x,t) \in \overline{\Omega}$ it is assumed that the components $a_{ij}(x,t)$ of $A(x,t)$ satisfy the inequalities

$$a_{ii}(x,t) > \sum_{\substack{j \ne i \\ j=1}}^{n} |a_{ij}(x,t)| \text{ for } 1 \le i \le n, \text{ and } a_{ij}(x,t) \le 0 \text{ for } i \ne j \tag{2}$$

and

$$0 < \alpha < \min_{\substack{(x,t)\in\overline{\Omega} \\ 1 \le i \le n}} \left(\sum_{j=1}^{n} a_{ij}(x,t)\right), \qquad \text{for some } \alpha. \tag{3}$$

It is also assumed that $\sqrt{\varepsilon_n} \le \dfrac{\sqrt{\alpha}}{6}$. Further $\mathbf{f}$ and A are assumed to be sufficiently smooth and sufficient compatibility conditions are assumed such that $\mathbf{u} \in C_\lambda^6(\overline{\Omega})$, for $A, \mathbf{f} \in C_\lambda^3(\overline{\Omega})$. Here

$$C_\lambda^k(D) = \{u : \frac{\partial^{l+m} u}{\partial x^l \partial t^m} \in C_\lambda^0(D) \text{ for } l, m \ge 0,\ l \le 4, \text{ and } 0 \le l + 2m \le k\}.$$

The reduced problem corresponding to (1) is defined by

$$\frac{\partial \mathbf{u}_0}{\partial t} + A\mathbf{u}_0 = \mathbf{f}, \text{ on } \Omega, \quad \mathbf{u}_0 = \mathbf{u} \text{ on } \{(x,0) : 0 < x < 1\}.$$

For any vector-valued function $\mathbf{y}$ on $\overline{\Omega}$ the following norms are introduced:
$\| \mathbf{y}(x,t) \| = \max_i |y_i(x,t)|$ and $\| \mathbf{y} \| = \sup\{\| \mathbf{y}(x,t) \| : (x,t) \in \overline{\Omega}\}$ Throughout the paper C denotes a generic positive constant, which is independent of x, t and of all singular perturbation and discretization parameters. Furthermore, inequalities between vectors are understood in the componentwise sense. Whenever necessary the required smoothness of the problem data is assumed.

For a general introduction to parameter-uniform numerical methods for singular perturbation problems, see [1–3] and for parameter-uniform numerical methods for singularly perturbed parabolic problems, see [4–8]. In [7] the general $n \times n$ system is considered and uniform convergence of first order in time and essentially first order in space is proved. In [8], for the problem under consideration, first order convergence in time and essentially second order convergence in space is established. In [5], (1) is considered in the special case $n = 2$. A numerical method combining the Crank-Nicolson operator in time and a central difference operator in space on a piecewise uniform Shishkin mesh is used. Second order convergence in time and essentially second order convergence in space is proved under the additional restrictions $\sqrt{\varepsilon_1} <$

$\sqrt{\varepsilon_2} \leq N^{-1}$. In the present paper these results are generalised to an $n \times n$ system. Moreover, no restrictions of the above kind on the ε_i and N are required. These results are made possible by the use of a sequence of layer interaction points, which were introduced in [8]. A single layer interaction point was used in [6] in the special case $n = 2$ and $s = 1$. Note that, as in [5], the power-boundedness of a family of operators is assumed.

The plan of the paper is as follows. In Sect. 2 estimates of the analytical behaviour of the exact solution are presented without proof; these are available in [8]. In Sect. 3, the Crank-Nicolson semi-discretisation in time is defined and the error is estimated. In Sect. 4, the complete discretisation in time and space is introduced. The central difference scheme is used on a piecewise uniform Shishkin mesh for the spatial discretisation. In Sect. 5, the error of this complete discretisation is estimated.

2 Analytical Estimates of the Exact Solution

The proofs of all the lemmas in this section may be found in [8]. The operator $\mathbf{L}$ satisfies the following maximum principle

Lemma 1 *Let $A(x,t)$ satisfy (2) and (3). Let $\boldsymbol{\psi}$ be any vector-valued function in the domain of $\mathbf{L}$ such that $\boldsymbol{\psi} \geq \mathbf{0}$ on Γ. Then $\mathbf{L}\boldsymbol{\psi}(x,t) \geq \mathbf{0}$ on Ω implies that $\boldsymbol{\psi}(x,t) \geq \mathbf{0}$ on $\overline{\Omega}$.*

Lemma 2 *Let $A(x,t)$ satisfy (2) and (3). If $\boldsymbol{\psi}$ is any vector-valued function in the domain of $\mathbf{L}$, then, for each i, $1 \leq i \leq n$ and $(x,t) \in \overline{\Omega}$,*

$$|\psi_i(x,t)| \leq \max\left\{\| \boldsymbol{\psi} \|_{\Gamma}, \frac{1}{\alpha} \| \mathbf{L}\boldsymbol{\psi} \|\right\}.$$

The Shishkin decomposition of the exact solution $\mathbf{u}$ of (1) is $\mathbf{u} = \mathbf{v} + \mathbf{w}$ where the smooth component $\mathbf{v}$ is the solution of $\mathbf{Lv} = \mathbf{f}$ in Ω, $\mathbf{v} = \mathbf{u}_0$ on Γ and the singular component $\mathbf{w}$ is the solution of $\mathbf{Lw} = \mathbf{0}$ in Ω, $\mathbf{w} = \mathbf{u} - \mathbf{v}$ on Γ. For convenience the left and right boundary layers of $\mathbf{w}$ are separated using the further decomposition $\mathbf{w} = \mathbf{w}^L + \mathbf{w}^R$ where $\mathbf{Lw}^L = \mathbf{0}$ on Ω, $\mathbf{w}^L = \mathbf{w}$ on Γ_L, $\mathbf{w}^L = \mathbf{0}$ on $\Gamma_B \cup \Gamma_R$ and $\mathbf{Lw}^R = \mathbf{0}$ on Ω, $\mathbf{w}^R = \mathbf{w}$ on Γ_R, $\mathbf{w}^R = \mathbf{0}$ on $\Gamma_L \cup \Gamma_B$.

Sharper estimates of smooth and singular components of the solution $\mathbf{u}$ are obtained by defining layer functions B_i^L, B_i^R, B_i, $i = 1, \ldots, n$, as follows.

$$B_i^L(x) = e^{-x\sqrt{\alpha/\varepsilon_i}}, \; B_i^R(x) = B_i^L(1-x), \; B_i(x) = B_i^L(x) + B_i^R(x).$$

The following elementary properties of these layer functions, for all $1 \leq i < j \leq n$ and $0 \leq x < y \leq 1$, should be noted:

$B_i(x) = B_i(1-x)$. $B_i^L(x) < B_j^L(x)$, $B_i^L(x) > B_i^L(y)$, $0 < B_i^L(x) \leq 1$.
$B_i^R(x) < B_j^R(x)$, $B_i^R(x) < B_i^R(y)$, $0 < B_i^R(x) \leq 1$.

$B_i(x)$ is monotone decreasing for increasing $x \in [0, \frac{1}{2}]$.
$B_i(x)$ is monotone increasing for increasing $x \in [\frac{1}{2}, 1]$.

$$B_i(x) \le 2B_i^L(x) \text{ for } x \in [0, \frac{1}{2}],\ B_i(x) \le 2B_i^R(x) \text{ for } x \in [\frac{1}{2}, 1].$$

$$B_i^L(2\sqrt{\frac{\varepsilon_i}{\alpha}} \ln N) = N^{-2}.$$

The layer interaction points $x_{i,j}^{(s)}$ are now defined.

Definition 1 For B_i^L, B_j^L, each i, j, $1 \le i \neq j \le n$ and each $s, s > 0$, the point $x_{i,j}^{(s)}$ is defined by

$$\frac{B_i^L(x_{i,j}^{(s)})}{\varepsilon_i^s} = \frac{B_j^L(x_{i,j}^{(s)})}{\varepsilon_j^s}.$$

It is remarked that $\frac{B_i^R(1-x_{i,j}^{(s)})}{\varepsilon_i^s} = \frac{B_j^R(1-x_{i,j}^{(s)})}{\varepsilon_j^s}$.

In the next lemma the existence and uniqueness of the points $x_{i,j}^{(s)}$ are shown.

Lemma 3 *For all i, j, such that $1 \le i < j \le n$ and $0 < s \le 3/2$, the points $x_{i,j}$ exist, are uniquely defined and satisfy the following inequalities*

$$\frac{B_i^L(x)}{\varepsilon_i^s} > \frac{B_j^L(x)}{\varepsilon_j^s},\quad x \in [0, x_{i,j}^{(s)}),\quad \frac{B_i^L(x)}{\varepsilon_i^s} < \frac{B_j^L(x)}{\varepsilon_j^s},\quad x \in (x_{i,j}^{(s)}, 1].$$

Moreover

$$x_{i,j}^{(s)} < x_{i+1,j}^{(s)},\ \text{if}\ i+1 < j\ \text{and}\ x_{i,j}^{(s)} < x_{i,j+1}^{(s)},\ \text{if}\ i < j.$$

Also

$$x_{i,j}^{(s)} < 2s\sqrt{\frac{\varepsilon_j}{\alpha}}\ \textit{and}\ x_{i,j}^{(s)} \in (0, \frac{1}{2})\ \text{if}\ i < j.$$

Analogous results hold for the B_i^R, B_j^R and the points $1 - x_{i,j}^{(s)}$.

In the following lemma estimates of the smooth component are presented.

Lemma 4 *Let $A(x,t)$ satisfy (2) and (3). Then the smooth component $\mathbf{v}$ of the solution $\mathbf{u}$ of (1) satisfies, for all $i = 1, \cdots, n$ and all $(x,t) \in \overline{\Omega}$,*

$$\begin{aligned} &|\frac{\partial^l v_i}{\partial x^l}(x,t)| \le C(1 + \sum_{q=i}^{n} \frac{B_q(x)}{\varepsilon_q^{\frac{l}{2}-1}}) \textit{for } l = 0, 1, 2, 3 \\ &|\frac{\partial^l v_i}{\partial x^{l-1}\partial t}(x,t)| \le C \textit{ for } l = 2, 3. \end{aligned}$$

Bounds on $\mathbf{w}^L$, $\mathbf{w}^R$ and their derivatives are contained in

Lemma 5 *Let $A(x,t)$ satisfy (2) and (3). Then there exists a constant C, such that, for each $(x,t) \in \overline{\Omega}$ and $i = 1, \ldots, n$,*

$$|\frac{\partial^l w_i^L}{\partial t^l}(x,t)| \le C B_n^L(x), \quad \text{for } l = 0, 1, 2, \quad |\frac{\partial^l w_i^L}{\partial x^l}(x,t)| \le C \sum_{q=i}^{n} \frac{B_q^L(x)}{\varepsilon_q^{\frac{l}{2}}}, \text{ for } l = 1, 2$$

$$|\frac{\partial^3 w_i^L}{\partial x^3}(x,t)| \le C \sum_{q=1}^n \frac{B_q^L(x)}{\varepsilon_q^{\frac{3}{2}}}, \quad |\frac{\partial^4 w_i^L}{\partial x^4}(x,t)| \le C \frac{1}{\varepsilon_i} \sum_{q=1}^n \frac{B_q^L(x)}{\varepsilon_q}.$$

Analogous results hold for the w_i^R and their derivatives.

3 Crank-Nicolson Semi-discretization in Time

On $[0, T]$, a uniform mesh with M mesh intervals, given by $\overline{\Omega}_t^M = \{k\Delta t, 0 \le k \le M, \Delta t = T/M\}$ is considered. The following Crank-Nicolson scheme is applied on this mesh

$$\begin{aligned} &\mathbf{u}^0(x) = \mathbf{u}(x,0), \\ &(I + \tfrac{\Delta t}{2}\mathbf{L}_x)\mathbf{u}^{k+1}(x) = \tfrac{\Delta t}{2}(\mathbf{f}^k + \mathbf{f}^{k+1})(x) + (I - \tfrac{\Delta t}{2}\mathbf{L}_x)\mathbf{u}^k(x), \\ &\mathbf{u}^{k+1}(0) = \mathbf{u}(0, t_{k+1}), \mathbf{u}^{k+1}(1) = \mathbf{u}(1, t_{k+1}), k = 0, \ldots, M-1 \end{aligned} \tag{4}$$

where $\mathbf{f}^k = \mathbf{f}(x, t_k)$, $k = 0, \ldots, M$.

It is helpful to introduce the following artificial problem:

$$\begin{aligned} &(I + \tfrac{\Delta t}{2}\mathbf{L}_x)\hat{\mathbf{u}}^{k+1}(x) = \tfrac{\Delta t}{2}(\mathbf{f}^k + \mathbf{f}^{k+1})(x) + (I - \tfrac{\Delta t}{2}\mathbf{L}_x)\mathbf{u}(x, t_k), \\ &\hat{\mathbf{u}}^{k+1}(0) = \mathbf{u}(0, t_{k+1}), \hat{\mathbf{u}}^{k+1}(1) = \mathbf{u}(1, t_{k+1}), \end{aligned} \tag{5}$$

where the exact solution $\mathbf{u}$ has replaced $\mathbf{u}^k$ in the right hand side of (4).

The operator $I + \frac{\Delta t}{2}\mathbf{L}_x$ satisfies the following maximum principle.

Lemma 6 *Let $A(x,t)$ satisfy (2) and (3). Let $\boldsymbol{\psi}$ be any vector-valued function in the domain of the operator $I + \frac{\Delta t}{2}\mathbf{L}_x$ such that $\boldsymbol{\psi}(0) \ge \mathbf{0}$ and $\boldsymbol{\psi}(1) \ge \mathbf{0}$. Then $(I + \frac{\Delta t}{2}\mathbf{L}_x)\boldsymbol{\psi}(x) \ge \mathbf{0}$ on $(0,1)$ implies that $\boldsymbol{\psi}(x) \ge \mathbf{0}$ on $[0,1]$.*

Proof Let $\psi_{i^*}(x^*) = \min_i \min_{[0,1]} \psi_i(x)$. Assume that $\psi_{i^*}(x^*) < 0$. Then $\frac{d^2\psi_{i^*}}{dx^2}(x^*) \ge 0$. From the hypotheses we have $x^* \notin \{0, 1\}$. Using (2), (3) it follows that $(I + \frac{\Delta t}{2}\mathbf{L}_x)_{i^*}\boldsymbol{\psi}(x^*) < 0$,

The stability of the operator $I + \frac{\Delta t}{2}\mathbf{L}_x$ is now established.

Lemma 7 *Let $A(x,t)$ satisfy (2) and (3). If $\boldsymbol{\psi}$ is any vector-valued function in the domain of the operator $I+\frac{\Delta t}{2}\mathbf{L}_x$ then, for each i, $1\le i\le n$ and $x\in[0,1]$,*

$$||\boldsymbol{\psi}(x)|| \le \max\left\{\|\boldsymbol{\psi}(0)\|, \|\boldsymbol{\psi}(1)\|, \frac{1}{\alpha}\|(I+\frac{\Delta t}{2}\mathbf{L}_x)\boldsymbol{\psi}(x)\|\right\}.$$

Proof Define the barrier functions

$$\boldsymbol{\theta}^{\pm}(x) = \max\left\{||\boldsymbol{\psi}(0)||, ||\boldsymbol{\psi}(1)||, \frac{1}{\alpha}||(I+\frac{\Delta t}{2}\mathbf{L}_x)\boldsymbol{\psi}(x)||\right\}\mathbf{e} \pm \boldsymbol{\psi}(x)$$

where $\mathbf{e}=(1,\ldots,1)^T$ is the unit column vector. Then $\boldsymbol{\theta}^{\pm}(x)\ge\mathbf{0}$ at $x=0,1$ and $(I+\frac{\Delta t}{2}\mathbf{L}_x)\boldsymbol{\theta}^{\pm}(x)\ge\mathbf{0}$ on $(0,1)$. The result follows from Lemma 6.

The error in the solution $\hat{\mathbf{u}}^{k+1}$ of the artificial problem is estimated in the following lemma.

Lemma 8 *If $\mathbf{u}(x,t)$ is the solution of (1) and $\hat{\mathbf{u}}^{k+1}(x)$ is the solution of (5), then, for $x\in[0,1]$ and $k=0,\ldots,M-1$,*

$$||\mathbf{u}(x,t_{k+1})-\hat{\mathbf{u}}^{k+1}(x)|| \le C(\Delta t)^3.$$

Proof Since $f^{k+\frac{1}{2}}(x)=\dfrac{f^{k+1}(x)+f^k(x)}{2}+\mathbb{O}((\Delta t)^2)$, from (1)

$$\frac{\partial\mathbf{u}(x,t_{k+\frac{1}{2}})}{\partial t}+\mathbf{L}_x\mathbf{u}(x,t_{k+\frac{1}{2}}) = \frac{\frac{\partial\mathbf{u}(x,t_{k+1})}{\partial t}+\mathbf{L}_x\mathbf{u}(x,t_{k+1})+\frac{\partial\mathbf{u}(x,t_k)}{\partial t}+\mathbf{L}_x\mathbf{u}(x,t_k)}{2} +\mathbb{O}((\Delta t)^2).$$

So

$$\mathbf{L}_x\mathbf{u}(x,t_{k+\frac{1}{2}}) = \mathbf{L}_x\frac{\mathbf{u}(x,t_{k+1})+\mathbf{u}(x,t_k)}{2}+\mathbb{O}((\Delta t)^2). \tag{6}$$

From (1),

$$\frac{\partial\mathbf{u}}{\partial t}(x,t_{k+1})+\mathbf{L}_x\mathbf{u}(x,t_{k+1}) = \mathbf{f}^{k+1}(x)$$

$$(\frac{\Delta t}{2})\mathbf{L}_x\mathbf{u}(x,t_{k+1}) = (\frac{\Delta t}{2})(-\frac{\partial\mathbf{u}}{\partial t}(x,t_{k+1})+\mathbf{f}^{k+1}(x))$$

$$(I+\frac{\Delta t}{2}\mathbf{L}_x)\mathbf{u}(x,t_{k+1}) = \mathbf{u}(x,t_{k+1})+\frac{\Delta t}{2}(\frac{-\partial\mathbf{u}}{\partial t}(x,t_{k+1})+\mathbf{f}^{k+1}(x)). \tag{7}$$

Then (7) and (5) yield

$$\begin{aligned}
(I + \tfrac{\Delta t}{2}\mathbf{L}_x)(\mathbf{u}(x,t_{k+1}) - \hat{\mathbf{u}}^{k+1}(x)) &= \mathbf{u}(x,t_{k+1}) + \tfrac{\Delta t}{2}(\tfrac{-\partial \mathbf{u}}{\partial t}(x,t_{k+1})) \\
&\quad - \tfrac{\Delta t}{2}\mathbf{f}^k(x) - (I - \tfrac{\Delta t}{2}\mathbf{L}_x)\mathbf{u}(x,t_k) \\
&= (\mathbf{u}(x,t_{k+1}) - \mathbf{u}(x,t_k)) + \tfrac{\Delta t}{2}(\mathbf{f}^{k+1}(x) + \mathbf{L}_x\mathbf{u}(x,t_{k+1})) + \tfrac{\Delta t}{2}(\mathbf{L}_x\mathbf{u}(x,t_k)) - \tfrac{\Delta t}{2}\mathbf{f}^k(x) \\
&= \tfrac{\Delta t}{2}\mathbf{L}_x(\mathbf{u}(x,t_{k+1}) + \mathbf{u}(x,t_k)) - \tfrac{\Delta t}{2}(\mathbf{f}^k + \mathbf{f}^{k+1})(x) + (\mathbf{u}(x,t_{k+1}) - \mathbf{u}(x,t_k)) \\
&= \Delta t\mathbf{L}_x(\mathbf{u}(x,t_{k+\frac{1}{2}}) + \mathbb{O}((\Delta t)^2)) - \Delta t(\mathbf{f}^{k+\frac{1}{2}}(x) + \mathbb{O}((\Delta t)^2)) \\
&\quad + \Delta t(\frac{\partial \mathbf{u}(x,t_{k+\frac{1}{2}})}{\partial t} + \mathbb{O}((\Delta t)^2)) \\
&= \mathbb{O}((\Delta t)^3), \text{ using (6).}
\end{aligned}$$

Then by Lemma 7 it follows that

$$||\mathbf{u}(x,t_{k+1}) - \hat{\mathbf{u}}^{k+1}(x)|| \le \mathbb{O}((\Delta t)^3).$$

Consider now the Shishkin decomposition of $\hat{\mathbf{u}}^{k+1}$ into smooth and singular components given by

$$\hat{\mathbf{u}}^{k+1} = \hat{\mathbf{v}}^{k+1} + \hat{\mathbf{w}}^{k+1},$$

where $\hat{\mathbf{v}}^{k+1}$ and $\hat{\mathbf{w}}^{k+1}$ are defined to be the solutions of the problems

$$\left.\begin{aligned}
(I + \tfrac{\Delta t}{2}\mathbf{L}_x)\hat{\mathbf{v}}^{k+1}(x) &= \tfrac{\Delta t}{2}(\mathbf{f}^k + \mathbf{f}^{k+1})(x) + (I - \tfrac{\Delta t}{2}\mathbf{L}_x)\mathbf{v}(x,t_k), \\
(I + \tfrac{\Delta t}{2}\mathbf{A})\hat{\mathbf{v}}^{k+1}(x) &= \tfrac{\Delta t}{2}(\mathbf{f}^k + \mathbf{f}^{k+1})(x) + (I - \tfrac{\Delta t}{2}\mathbf{A})\mathbf{v}(x,t_k), x = 0, 1
\end{aligned}\right\} \quad (8)$$

and

$$\left.\begin{aligned}
&(I + \tfrac{\Delta t}{2}\mathbf{L}_x)\hat{\mathbf{w}}^{k+1} = (I - \tfrac{\Delta t}{2}\mathbf{L}_x)\mathbf{w}(x,t_k), \\
&\hat{\mathbf{w}}^{k+1}(0) = \hat{\mathbf{u}}^{k+1}(0) - \hat{\mathbf{v}}^{k+1}(0),\ \hat{\mathbf{w}}^{k+1}(1) = \hat{\mathbf{u}}^{k+1}(1) - \hat{\mathbf{v}}^{k+1}(1).
\end{aligned}\right\} \quad (9)$$

The bounds on $\hat{\mathbf{v}}^{k+1}$ and its x-derivatives are contained in the following lemma.

Lemma 9 *The smooth component $\hat{\mathbf{v}}^{k+1}(x)$ and its derivatives satisfy, for each $i = 1, \ldots, n$ and $x \in [0, 1]$,*

$$|\frac{d^l \hat{v}_i^{k+1}}{dx^l}(x)| \le C(1 + \varepsilon_i^{1-\frac{l}{2}}) \qquad \text{for } l = 0, 1, 2, 3, 4.$$

Proof By the ith equation of the defining equations for $\hat{\mathbf{v}}^{k+1}(x)$,

$$((I + \frac{\Delta t}{2}\mathbf{L}_x)\hat{\mathbf{v}}^{k+1})_i(x) = \frac{\Delta t}{2}(f_i^k + f_i^{k+1})(x) + ((I - \frac{\Delta t}{2}\mathbf{L}_x)\mathbf{v})_i(x,t_k).$$

Using the bounds of v_i and its derivatives, $|((I+\frac{\Delta t}{2}\mathbf{L}_x)\hat{\mathbf{v}}^{k+1})_i(x)| \le C(1+\varepsilon_i)$.

Using Lemma 7, $|\hat{v_i}^{k+1}(x)| \le C(1+\varepsilon_i)$
Differentiating the ith equation of (8) twice with respect to x,

$$|((I+\frac{\Delta t}{2}\mathbf{L}_x)\frac{d^2\hat{\mathbf{v}}^{k+1}}{dx^2})_i(x)| \le C(I+\frac{d\hat{v}_i^{k+1}}{dx}(x)). \tag{10}$$

Let $|\frac{d\hat{v}_{i^*}^{k+1}}{dx}(x^*)| = ||\frac{d\hat{\mathbf{v}}^{k+1}}{dx}||$, for some $i=i^*, x=x^*$.

Then for some $y \in [0, 1-x^*]$, and for some $\theta \in (x^*, x^*+y)$, by the mean value theorem,

$$|\frac{d\hat{v}_{i^*}^{k+1}}{dx}(x^*)| \le \frac{2}{y}||\hat{\mathbf{v}}^{k+1}(x)|| + \frac{y}{2}||\frac{d^2\hat{\mathbf{v}}^{k+1}}{dx^2}(x)||. \tag{11}$$

Using (11) in (10)

$$||\frac{d^2\hat{\mathbf{v}}^{k+1}}{dx^2}(x)|| \le 1 + \frac{2}{y}||\hat{\mathbf{v}}^{k+1}(x)|| + \frac{y}{2}||\frac{d^2\hat{\mathbf{v}}^{k+1}}{dx^2}(x)||$$

or

$$(1-\frac{y}{2})||\frac{d^2\hat{\mathbf{v}}^{k+1}}{dx^2}(x)|| \le C(1+\frac{2}{y}||\hat{\mathbf{v}}^{k+1}(x)||).$$

Using the expression in (11) gives

$$||\frac{d\hat{\mathbf{v}}^{k+1}}{dx}(x)|| \le \frac{2}{y}||\hat{\mathbf{v}}^{k+1}(x)|| + C + \frac{C}{y}||\hat{\mathbf{v}}^{k+1}(x)|| \le C.$$

The bounds on $||\frac{d^3\hat{\mathbf{v}}^{k+1}}{dx^3}(x)||$, $||\frac{d^4\hat{\mathbf{v}}^{k+1}}{dx^4}(x)||$ can be obtained analogously.

For convenience the left and right boundary layers of $\hat{\mathbf{w}}^{k+1}$ are separated using the further decomposition $\hat{\mathbf{w}}^{k+1} = \hat{\mathbf{w}}^{k+1,L} + \hat{\mathbf{w}}^{k+1,R}$
where

$$\left.\begin{aligned} &(I+\tfrac{\Delta t}{2}\mathbf{L}_x)\hat{\mathbf{w}}^{k+1,L}(x) = (I-\tfrac{\Delta t}{2}\mathbf{L}_x)\mathbf{w}^L(x,t_k),\\ &\hat{\mathbf{w}}^{k+1,L}(0) = \hat{\mathbf{w}}^{k+1}(0),\ \hat{\mathbf{w}}^{k+1,L}(1) = \mathbf{0} \end{aligned}\right\} \tag{12}$$

and

$$\left.\begin{aligned} &(I+\tfrac{\Delta t}{2}\mathbf{L}_x)\hat{\mathbf{w}}^{k+1,R}(x) = (I-\tfrac{\Delta t}{2}\mathbf{L}_x)\mathbf{w}^R(x,t_k),\\ &\hat{\mathbf{w}}^{k+1,R}(0) = \mathbf{0},\ \hat{\mathbf{w}}^{k+1,R}(1) = \hat{\mathbf{w}}^{k+1}(1). \end{aligned}\right\}$$

Bounds on the singular components $\hat{\mathbf{w}}^{k+1,L}$, $\hat{\mathbf{w}}^{k+1,R}$ of $\hat{\mathbf{u}}^{k+1}$ and their x-derivatives are contained in the next lemma.

Lemma 10 *For $i = 1, \ldots, n$ and $x \in [0, 1]$, the singular component $\hat{\mathbf{w}}^{k+1,L}(x)$ and its derivatives satisfy*

$$|\hat{w}_i^{k+1,L}(x)| \le CB_n^L(x), \quad |\frac{d^l \hat{w}_i^{k+1,L}}{dx^l}(x)| \le C\sum_{q=i}^n \frac{B_q^L(x)}{\varepsilon_q^{\frac{l}{2}}}, \text{ for } l = 1, 2.$$

$$|\frac{d^3 \hat{w}_i^{k+1,L}}{dx^3}(x)| \le C\sum_{q=1}^n \frac{B_q^L(x)}{\varepsilon_q^{\frac{3}{2}}} \text{ and } |\frac{d^4 \hat{w}_i^{k+1,L}}{dx^4}(x)| \le C\frac{1}{\varepsilon_i}\sum_{q=1}^n \frac{B_q^L(x)}{\varepsilon_q}.$$

Analogous results hold for the $\hat{\mathbf{w}}^{k+1,R}$ and its derivatives.

Proof Considering the ith equation of (12),

$$|((I + \frac{\Delta t}{2}\mathbf{L}_x)\hat{\mathbf{w}}^{k+1,L})_i(x)| = |((I - \frac{\Delta t}{2}\mathbf{L}_x)\mathbf{w}^L)_i(x, t_k)|.$$

Then, using the fact that $(\mathbf{L}_x\mathbf{w}^L)_i(x, t_k) = -\frac{\partial}{\partial t}w_i^L(x, t_k)$ and Lemma 5, it follows that $|((I + \frac{\Delta t}{2}\mathbf{L}_x)\hat{\mathbf{w}}^{k+1,L})_i(x)| \le CB_n^L(x)$.

Then by Lemma 7, $|\hat{w}_i^{k+1,L}(x)| \le CB_n^L(x)$.

Differentiating the ith equation of (12) partially with respect to x, using Lemma 5 and the bound for $\hat{w}_j^{k+1,L}$, it follows that

$$\begin{aligned}|((I + \tfrac{\Delta t}{2}\mathbf{L}_x)\tfrac{d\hat{\mathbf{w}}^{k+1,L}}{dx})_i(x)| \le\ & CB_n^L(x) + C\textstyle\sum_{q=i}^n \frac{B_q^L}{\sqrt{\varepsilon_i}}(x) \\ & + \tfrac{\Delta t}{2}|(\mathbf{L}_x\tfrac{\partial \mathbf{w}^L}{\partial x})_i(x, t_k)| + CB_n^L(x).\end{aligned} \tag{13}$$

It can be found in the proof of Lemma 4.3 in [8] that

$$|\frac{\partial^{l+m} w_i^L}{\partial x^l \partial t^m}(x)| \le C\varepsilon_i^{\frac{-l}{2}} B_n^L(x), \quad l \le 3, \; m \le 2 \text{ and } 0 \le l + 2m \le 4. \tag{14}$$

Substituting the bound for $\frac{\partial^2 w_i^L}{\partial x \partial t}(x, t_k)$ in

$$|(\mathbf{L}_x\frac{\partial \mathbf{w}^L}{\partial x})_i(x, t_k)| \le |\frac{\partial^2 w_i^L}{\partial x \partial t}(x, t_k)| + |\sum_{j=1}^n \frac{\partial a_{ij}}{\partial x}(x, t_k)w_j^L(x, t_k)|$$

yields

$$|(\mathbf{L}_x\frac{\partial \mathbf{w}^L}{\partial x})_i(x, t_k)| \le C\sum_{q=i}^n \frac{B_q^L}{\sqrt{\varepsilon_q}}(x) + CB_n^L(x). \tag{15}$$

Using (15) in (13) and Lemma 7, the bound $|\frac{d\hat{w}_i^{k+1,L}}{dx}(x)| \le C\sum_{q=i}^{n} \frac{B_q^L}{\sqrt{\varepsilon_q}}(x)$ follows. Differentiating the ith equation of (12) twice with respect to x, and using (14), Lemma 5 and the bounds of $\hat{w}_i^{k+1,L}$, $\frac{d\hat{w}_i^{k+1,L}}{dx}(x)$ gives

$|((I + \frac{\Delta t}{2}L_x)\frac{d^2\hat{\mathbf{w}}^{k+1,L}}{dx^2})_i(x)| \le \sum_{q=i}^{n} \frac{B_q^L(x)}{\varepsilon_q^1}$.

Then, using Lemma 7, the bound on $|\frac{d^2\hat{w}_i^{k+1,L}}{dx^2}(x)|$ follows.

Using the appropriate barrier functions and the techniques used in [8], it follows that $|\frac{\partial^4 w_i^L}{\partial x^3 \partial t}(x)| \le C\varepsilon_i^{\frac{-3}{2}} B_n^L(x)$ and $|\frac{\partial^5 w_i^L}{\partial x^4 \partial t}(x)| \le C\varepsilon_i^{-2} B_n^L(x)$. Then, the bounds on $|\dfrac{d^3\hat{w}_i^{k+1,L}}{dx^3}(x)|$ and $|\dfrac{d^4\hat{w}_i^{k+1,L}}{dx^4}(x)|$ are obtained by repeating the same procedure.

A similar proof of the analogous results for the boundary layer functions $\hat{\mathbf{w}}^{k+1,R}$ holds.

Sharper bounds on $\hat{\mathbf{v}}^{k+1}$ and its derivatives are contained in the following lemma

Lemma 11 *The smooth component* $\hat{\mathbf{v}}^{k+1}(x)$ *and its* x*-derivatives satisfy, for each* $i = 1, \ldots, n$ *and* $x \in [0, 1]$,

$$|\frac{d^l \hat{v}_i^{k+1}}{dx^l}(x)| \le C(1 + \sum_{q=i}^{n} \frac{B_q(x)}{\varepsilon_q^{\frac{l}{2}-1}}) \qquad \text{for} \qquad l = 0, 1, 2, 3.$$

Proof Define the barrier functions $\boldsymbol{\psi}^{\pm}(x) = C[1 + B_n(x)]\mathbf{e} \pm \frac{d^l\hat{\mathbf{v}}^{k+1}}{dx^l}(x)$, $l = 0, 1, 2$, $x \in [0, 1]$. It follows from Lemma 6 that $|\frac{d^l\hat{v}_i^{k+1}}{dx^l}(x)| \le C[1 + B_n(x)]$, $l = 0, 1, 2$. Let $\mathbf{p} = \frac{d^2\hat{\mathbf{v}}^{k+1}}{dx^2}(x)$, then

$$(I + \frac{\Delta t}{2}\mathbf{L}_x)\mathbf{p}(x) = \mathbf{g}(x, t) \text{ with } \mathbf{p}(0) = \mathbf{0},\ \mathbf{p}(1) = 0, \tag{16}$$

where

$$\begin{aligned} \mathbf{g}(x, t) = \frac{\Delta t}{2}(\frac{\partial^2 \mathbf{f}^k}{\partial x^2} + \frac{\partial^2 \mathbf{f}^{k+1}}{\partial x^2})(x) + (I - \frac{\Delta t}{2}\mathbf{L}_x)\frac{\partial^2}{\partial x^2}\mathbf{v}(x, t_k) - 2\frac{\partial A}{\partial x}\frac{d\hat{\mathbf{v}}^{k+1}}{dx}(x) \\ + 2\frac{\partial A}{\partial x}\frac{\partial \mathbf{v}}{\partial x}(x, t_k) - \frac{\partial^2 A}{\partial x^2}\mathbf{v}(x, t_k) - \frac{d^2 A}{dx^2}\hat{\mathbf{v}}^{k+1}(x). \end{aligned} \tag{17}$$

Using the bound $|\frac{\partial^3}{\partial x^2 \partial t}\mathbf{v}(x, t_k)| \le C$ in the expression for $\mathbf{L}_x \frac{\partial^2}{\partial x^2}\mathbf{v}(x, t_k)$ leads to the bound $|\mathbf{L}_x \frac{\partial^2}{\partial x^2}\mathbf{v}(x, t_k)| \le C$ and it follows that $|\mathbf{g}(x, t)| \le C$.

Let $\mathbf{p} = \mathbf{q} + \mathbf{r}$, where $\mathbf{q}$ and $\mathbf{r}$, the smooth and singular components of $\mathbf{p}$, are defined to be the solutions of the following problems: $(I + \frac{\Delta t}{2}\mathbf{L}_x)\mathbf{q} = \mathbf{g}(x, t)$, with $\mathbf{q}(0) = \mathbf{p}_0(0)$, $\mathbf{q}(1) = \mathbf{p}_0(1)$, where $\mathbf{p}_0$ is the solution of the reduced problem of (16), and $(I + \frac{\Delta t}{2}\mathbf{L}_x)\mathbf{r} = \mathbf{0}$, with $\mathbf{r}(0) = -\mathbf{q}(0)$, $\mathbf{r}(1) = -\mathbf{q}(1)$.

By Lemma 9, $|\frac{dq_i}{dx}(x)| \le C$ and by Lemma 10, $|\frac{dr_i}{dx}(x)| \le C(1 + \sum_{q=i}^{n} \frac{B_q(x)}{\sqrt{\varepsilon_q}})$. Thus,

$$|\frac{dp_i}{dx}(x)| = |\frac{d^3\hat{v}_i^{k+1}}{dx^3}(x)| \le (1 + \sum_{q=i}^{n} \frac{B_q(x)}{\sqrt{\varepsilon_q}}),$$

which completes the proof of the lemma.

4 The Complete Discretisation in Time and Space

A piecewise uniform Shishkin mesh with $M \times N$ mesh-intervals is now constructed. Let $\Omega_t^M = \{t_k\}_{k=1}^M$, $\Omega_x^N = \{x_j\}_{j=1}^{N-1}$, $\overline{\Omega}_t^M = \{t_k\}_{k=0}^M$, $\overline{\Omega}_x^N = \{x_j\}_{j=0}^N$, $\Omega^{M,N} = \Omega_t^M \times \Omega_x^N$, $\overline{\Omega}^{M,N} = \overline{\Omega}_t^M \times \overline{\Omega}_x^N$ and $\Gamma^{M,N} = \Gamma \cap \overline{\Omega}^{M,N}$. The mesh $\overline{\Omega}_t^M$ is chosen to be a uniform mesh with M mesh-intervals on $[0, T]$. Let $\{\varepsilon_r\}_1^{n'}$ be the set of all distinct parameters in $\{\varepsilon_r\}_1^n$. The mesh $\overline{\Omega}_x^N$ is a piecewise-uniform mesh on $[0, 1]$ obtained by dividing $[0, 1]$ into $2n' + 1$ mesh-intervals as follows

$$[0, \sigma_1] \cup \cdots \cup (\sigma_{n'-1}, \sigma_{n'}] \cup (\sigma_{n'}, 1 - \sigma_{n'}] \cup (1 - \sigma_{n'}, 1 - \sigma_{n'-1}] \cup \cdots \cup (1 - \sigma_1, 1].$$

The n' parameters σ_r, which determine the points separating the uniform meshes, are defined by $\sigma_0 = 0$, $\sigma_{n'+1} = \frac{1}{2}$ and, for $r = 1, \ldots, n'$,

$$\sigma_r = \min\left\{\frac{\sigma_{r+1}}{2}, 2\sqrt{\frac{\varepsilon_r}{\alpha}} \ln N\right\}.$$

Clearly

$$0 < \sigma_1 < \ldots < \sigma_{n'} \le \frac{1}{4}, \quad \frac{3}{4} \le 1 - \sigma_{n'} < \ldots < 1 - \sigma_1 < 1.$$

Then, on the sub-interval $(\sigma_{n'}, 1 - \sigma_{n'}]$ a uniform mesh with $\frac{N}{2}$ mesh-intervals is placed, on each of the sub-intervals $(\sigma_r, \sigma_{r+1}]$ and $(1 - \sigma_{r+1}, 1 - \sigma_r]$, $r = 1, \ldots, n' - 1$, a uniform mesh of $\frac{N}{2^{n'-r+2}}$ mesh-intervals is placed and on both of the sub-intervals $[0, \sigma_1]$ and $(1 - \sigma_1, 1]$ a uniform mesh of $\frac{N}{2^{n'+1}}$ mesh-intervals is placed. In practice it is convenient to take

$$N = 2^{n'+p+1}$$

for some natural number p. It follows that, for $2 \le r \le n'$, in the sub-interval $[\sigma_{r-1}, \sigma_r]$ there are $N/2^{n'-r+3} = 2^{r+p-2}$ mesh-intervals and in each of $[0, \sigma_1]$, $[\sigma_1, \sigma_2]$, $[1 - \sigma_2, 1 - \sigma_1]$ and $[1 - \sigma_1, 1]$ there are $N/2^{n'+1} = 2^p$. This construction leads to a class of $2^{n'}$ piecewise uniform Shishkin meshes $\Omega^{M,N}$. Note that these meshes are not the same as those constructed in [7].

Lemma 12 *For some r, $1 \le r \le n'$, suppose that $\sigma_r < \frac{\sigma_{r+1}}{2}$. Then the following inequalities hold*

$$B_r^L(1-\sigma_r) \le B_r^L(\sigma_r) = N^{-2}.$$

$$x^{(s)}_{r-1,r} \;\le\; \sigma_r - h_r^- \text{ for } 0 < s \le 2, 1 < r \le n'.$$

$$B_q^L(\sigma_r - h_r^-) \le C B_q^L(\sigma_r) \text{ for } 1 \le r \le q \le n'.$$

$$\frac{B_q^L(\sigma_r)}{\sqrt{\varepsilon_q}} \le C\frac{1}{\sqrt{\varepsilon_r}\ln N} \text{ for } 1 \le q \le n', \;\; 1 \le r \le n'.$$

Analogous results hold for B_r^R.

Proof The proof is as given in [9].

The discrete initial-boundary value problem is now defined on any mesh by

$$\left.\begin{array}{l}\mathbf{U}^0(x_j) = \mathbf{u}(x_j,0) \text{ on } \Gamma_B^N,\\ (I + \frac{\Delta t}{2}\mathbf{L}_x^N)\mathbf{U}^{k+1}(x_j) = \frac{\Delta t}{2}(\mathbf{f}^k + \mathbf{f}^{k+1})(x_j) + (I - \frac{\Delta t}{2}\mathbf{L}_x^N)\mathbf{U}^k(x_j),\\ \mathbf{U}^{k+1}(x_j) = \mathbf{u}(x_j, t_{k+1}) \text{ on } \Gamma_L^M \cup \Gamma_R^M, \text{ for } k = 0, ..., M-1\end{array}\right\} \quad (18)$$

where

$$\mathbf{L}_x^N \;=\; -E\delta_x^2 + A$$

and δ_x^2, D_x^+ and D_x^- are the standard difference operators

$$\delta_x^2\mathbf{U}(x_j, t_{k+1}) \;=\; \frac{D_x^+\mathbf{U}(x_j,t_{k+1}) - D_x^-\mathbf{U}(x_j,t_{k+1})}{(x_{j+1}-x_{j-1})/2},$$

$$D_x^+\mathbf{U}(x_j, t_{k+1}) \;=\; \frac{\mathbf{U}(x_{j+1},t_{k+1}) - \mathbf{U}(x_j,t_{k+1})}{x_{j+1}-x_j},$$

$$D_x^-\mathbf{U}(x_j, t_{k+1}) \;=\; \frac{\mathbf{U}(x_j,t_{k+1}) - \mathbf{U}(x_{j-1},t_{k+1})}{x_j-x_{j-1}}.$$

The following results for the discrete operator $I + \frac{\Delta t}{2}\mathbf{L}_x^N$ are analogous to Lemmas 6 and 7 for the continuous case. They are presented below without proof.

Lemma 13 *Let $A(x,t)$ satisfy (2) and (3). Then, for any vector-valued mesh function $\mathbf{\Psi}$, the inequalities $\mathbf{\Psi} \ge \mathbf{0}$ on $\Gamma^{M,N}$ and $(I + \frac{\Delta t}{2}\mathbf{L}_x^N)\mathbf{\Psi} \ge \mathbf{0}$ on $\Omega^{M,N}$ imply that $\mathbf{\Psi} \ge \mathbf{0}$ on $\overline{\Omega}^{M,N}$.*

An immediate consequence of this is the following discrete stability result.

Lemma 14 *Let $A(x,t)$ satisfy (2) and (3). Then, for any vector-valued mesh function $\mathbf{\Psi}$ on $\overline{\Omega}^{M,N}$, for $k = 0, ..., M-1$ and $0 \le j \le N$,*

$$\| \boldsymbol{\Psi}(x_j, t_{k+1}) \| \le \max\left\{ ||\boldsymbol{\Psi}||_{\Gamma^{M,N}}, \frac{1}{\alpha}||(I + \frac{\Delta t}{2}\mathbf{L}_x^N)\boldsymbol{\Psi}|| \right\}.$$

The discrete analogue of the artificial problem (5) is now defined by

$$\left.\begin{array}{l} (I + \frac{\Delta t}{2}\mathbf{L}_x^N)\hat{\mathbf{u}}^{k+1}(x_j) = \frac{\Delta t}{2}(\mathbf{f}^k + \mathbf{f}^{k+1})(x_j) + (I - \frac{\Delta t}{2}\mathbf{L}_x^N)\mathbf{u}(x_j, t_k), \\ \hat{\mathbf{u}}^{k+1} = \hat{\mathbf{u}}^{k+1} \text{ at } x_j = 0, 1, \text{ for } 0 < j < N, \text{ and } k = 0, ..., M-1. \end{array}\right\} \tag{19}$$

and the decomposition of its solution by $\hat{\mathbf{U}}^{k+1} = \hat{\mathbf{V}}^{k+1}(x_j) + \hat{\mathbf{W}}^{k+1}(x_j)$, where $\hat{\mathbf{V}}^{k+1}(x_j)$, $\hat{\mathbf{W}}^{k+1}(x_j)$ are the discrete analogues of $\hat{\mathbf{v}}^{k+1}(x_j)$, $\hat{\mathbf{w}}^{k+1}(x_j)$, which are respectively, the solutions of the problems, for $k = 0, ..., M-1$,

$$\left.\begin{array}{l} (I + \frac{\Delta t}{2}\mathbf{L}_x^N)\hat{\mathbf{V}}^{k+1}(x_j) = \frac{\Delta t}{2}(\mathbf{f}^k + \mathbf{f}^{k+1})(x_j) + (I - \frac{\Delta t}{2}\mathbf{L}_x^N)\mathbf{v}(x_j, t_k), \\ \hat{\mathbf{V}}^{k+1}(0) = \hat{\mathbf{v}}^{k+1}(0), \hat{\mathbf{V}}^{k+1}(1) = \hat{\mathbf{v}}^{k+1}(1), 0 < j < N, \end{array}\right\}$$

and

$$\left.\begin{array}{l} (I + \frac{\Delta t}{2}\mathbf{L}_x^N)\hat{\mathbf{W}}^{k+1}(x_j) = \frac{\Delta t}{2}(\mathbf{f}^k + \mathbf{f}^{k+1})(x_j) + (I - \frac{\Delta t}{2}\mathbf{L}_x^N)\mathbf{w}(x_j, t_k), \\ \hat{\mathbf{W}}^{k+1}(0) = \hat{\mathbf{w}}^{k+1}(0), \hat{\mathbf{W}}^{k+1}(1) = \hat{\mathbf{w}}^{k+1}(1), 0 < j < N. \end{array}\right\}$$

5 Error Estimate

Lemma 15 *Let $\hat{\mathbf{V}}^{k+1}(x)$ be the solution of (5) and $\hat{\mathbf{U}}^{k+1}(x_j)$ be the solution of (19). Then*

$$||\hat{\mathbf{V}}^{k+1}(x_j) - \hat{\mathbf{u}}^{k+1}(x_j)|| \le C\Delta t(N^{-1}\ln N)^2.$$

Proof From (5) and (19),

$$\begin{aligned} (I + \tfrac{\Delta t}{2}\mathbf{L}_x^N)(\hat{\mathbf{u}}^{k+1}(x_j) - \hat{\mathbf{u}}^{k+1}(x_j)) = {} & \tfrac{\Delta t}{2}(\mathbf{L}_x - \mathbf{L}_x^N)\hat{\mathbf{u}}^{k+1}(x_j) \\ & + \tfrac{\Delta t}{2}(\mathbf{L}_x - \mathbf{L}_x^N)\mathbf{u}(x, t_k). \end{aligned} \tag{20}$$

From [8, 9] it is found that

$$||(\mathbf{L}_x - \mathbf{L}_x^N)\mathbf{u}(x, t_k)|| \le C(N^{-1}(\ln N))^2. \tag{21}$$

Now, using the Shishkin decomposition of $\hat{\mathbf{u}}^{k+1}$, and the arguments used in [8, 9] on the different segments of the Shishkin mesh, the following hold

$$||\frac{\Delta t}{2}(\mathbf{L}_x - \mathbf{L}_x^N)\hat{\mathbf{v}}^{k+1}(x_j)|| \le C\Delta t(N^{-1}(\ln N))^2,$$

$$||\frac{\Delta t}{2}(\mathbf{L}_x - \mathbf{L}_x^N)\hat{\mathbf{w}}^{k+1}(x_j)|| \le C\Delta t(N^{-1}(\ln N))^2.$$

Using the above two expressions and (21) in (20) and Lemma 14, the required result follows.

Lemma 16 *Let* $\mathbf{u}(x_j, t_k)$ *be the solution of (1) and* $\mathbf{U}^{k+1}(x_j)$ *be the solution of (18). Then*

$$\begin{aligned}\mathbf{u}(x_j, t_{k+1}) - \mathbf{U}^{k+1}(x_j) = \textstyle\sum_{q=0}^{k} R^q\{&[\mathbf{u}(x_j, t_{k+1-q}) - \hat{\mathbf{u}}^{k+1-q}(x_j)] \\ &+[\hat{\mathbf{u}}^{k+1-q}(x_j) - \hat{\mathbf{U}}^{k+1-q}(x_j)]\},\end{aligned}$$

where R *is the operator given by* $R = (I + \frac{\Delta t}{2}\mathbf{L}_x^N)^{-1}(I - \frac{\Delta t}{2}\mathbf{L}_x^N)$, *for* $j = 0, 1, ..., N$ *and* $k = 0, 1, ..., M-1$.

Proof Subtracting (18) and (19) gives $(I + \frac{\Delta t}{2}\mathbf{L}_x^N)(\hat{\mathbf{U}}^{k+1} - \mathbf{U}^{k+1})(x_j) = (I - \frac{\Delta t}{2}\mathbf{L}_x^N)(\mathbf{u}(x_j, t_k) - \mathbf{U}^k(x_j))$ and so

$$\hat{\mathbf{U}}^{k+1}(x_j) - \mathbf{U}^{k+1}(x_j) = R(\mathbf{u}(x_j, t_k) - \mathbf{U}^k(x_j))$$

It is clear that

$$\begin{aligned}\mathbf{u}(x_j, t_{k+1}) - \mathbf{U}^{k+1}(x_j) = &(\mathbf{u}(x_j, t_{k+1}) - \hat{\mathbf{u}}^{k+1}(x_j)) + (\hat{\mathbf{u}}^{k+1}(x_j) - \hat{\mathbf{U}}^{k+1}(x_j)) \\ &+(\hat{\mathbf{U}}^{k+1}(x_j) - \mathbf{U}^{k+1}(x_j)).\end{aligned}$$

Now,

$$\begin{aligned}\mathbf{u}(x_j, t_{k+1}) - \mathbf{U}^{k+1}(x_j) = &[\mathbf{u}(x_j, t_{k+1}) - \hat{\mathbf{u}}^{k+1}(x_j)] + [\hat{\mathbf{u}}^{k+1}(x_j) - \hat{\mathbf{U}}^{k+1}(x_j)] \\ &+R\{[\mathbf{u}(x_j, t_k) - \hat{\mathbf{U}}^{k}(x_j)] + [\hat{\mathbf{u}}^{k}(x_j) - \hat{\mathbf{U}}^{k}(x_j)] \\ &\qquad +R[\mathbf{u}(x_j, t_{k-1}) - \mathbf{U}^{k-1}(x_j)]\}.\end{aligned}$$

Iterating,

$$\begin{aligned}\mathbf{u}(x_j, t_{k+1}) - \mathbf{U}^{k+1}(x_j) = \textstyle\sum_{q=0}^{k} R^q\{&[\mathbf{u}(x_j, t_{k+1-q}) - \hat{\mathbf{U}}^{k+1-q}(x_j)] \\ &+[\hat{\mathbf{U}}^{k+1-q}(x_j) - \hat{\mathbf{U}}^{k+1-q}(x_j)]\}.\end{aligned}$$

Power-Bound of the operator R

Definition 2 The family of operators R is said to be power-bounded if $||R^l|| \le C$ for all integers $l \ge 0$ and for some constant C independent of M, N, i, j and $\boldsymbol{\varepsilon}$.

In the following the power-boundedness of the operator R is proved for the special case where $n = 1$, the co-efficient function $a(x, t)$ is a constant and the x-domain is $\mathscr{R}$.

The discrete problem corresponding to the Crank-Nicolson scheme can be restated as

$$(1+\frac{\Delta t}{2}(-\varepsilon\delta_x^2+a))U^k(x_j) = (1-\frac{\Delta t}{2}(-\varepsilon\delta_x^2+a))U^{k-1}(x_j)+\frac{\Delta t}{2}(f^k+f^{k-1})(x_j). \tag{22}$$

where

$$\delta_x^2 U^k(x_j) = \frac{2}{h^+-h^-}(\frac{U^k(x_{j+1})-U^k(x_j)}{h^+} - \frac{U^k(x_j)-U^k(x_{j-1})}{h^-}).$$

Using Fourier transformation on (22)

$$\begin{aligned}&(1+\tfrac{\Delta t}{2}(a+\tfrac{4\varepsilon}{h^+-h^-}(\tfrac{sin^2\frac{\omega h^+}{2}}{h^+}+\tfrac{sin^2\frac{\omega h^-}{2}}{h^-})-\tfrac{2i\varepsilon}{h^+-h^-}(sin\tfrac{\omega h^+}{h^+}-sin\tfrac{\omega h^-}{h^-}))\tilde{U}^k(\omega)\\ &=(1-\tfrac{\Delta t}{2}(a+\tfrac{4\varepsilon}{h^+-h^-}(\tfrac{sin^2\frac{\omega h^+}{2}}{h^+}+\tfrac{sin^2\frac{\omega h^-}{2}}{h^-})-\tfrac{2i\varepsilon}{h^+-h^-}(sin\tfrac{\omega h^+}{h^+}-sin\tfrac{\omega h^-}{h^-}))\tilde{U}^{k-1}(\omega)\\ &\quad+\tfrac{\Delta t}{2}(\tilde{f}^k+\tilde{f}^{k-1})(\omega)\end{aligned}$$

The above can be rewritten as

$$(1+L)\tilde{U}^k(\omega) = (1-L)\tilde{U}^{k-1}(\omega)+\frac{\Delta t}{2}(\tilde{f}^k+\tilde{f}^{k-1})(\omega) \tag{23}$$

where

$$L = \frac{\Delta t}{2}(a+\frac{4\varepsilon}{h^+-h^-}(\frac{sin^2\frac{\omega h^+}{2}}{h^+}+\frac{sin^2\frac{\omega h^-}{2}}{h^-})-\frac{2i\varepsilon}{h^+-h^-}(sin\frac{\omega h^+}{h^+}-sin\frac{\omega h^-}{h^-})).$$

In order that (23) produces solution $\tilde{U}^k(\omega)$ which is bounded it suffices that $|R| = |\dfrac{1-L}{1+L}| \le 1$.

Equivalently $\dfrac{1-L}{1+L}.\dfrac{1-\overline{L}}{1+\overline{L}} \le 1$ or $2(L+\overline{L}) \ge 0$.

But

$$L+\overline{L} = \Delta t(a+\frac{4\varepsilon}{h^+-h^-}(\frac{sin^2\frac{\omega h^+}{2}}{h^+}+\frac{sin^2\frac{\omega h^-}{2}}{h^-})) > 0$$

which concludes the proof.

In the case of a system of n equations where $n > 1$, the coefficient is a constant and the domain is $\mathscr{R}$, we have

$$L = \frac{\Delta t}{2}(A+\frac{4E}{h^+-h^-}(\frac{sin^2\frac{\omega h^+}{2}}{h^+}+\frac{sin^2\frac{\omega h^-}{2}}{h^-})-\frac{2iE}{h^+-h^-}(sin\frac{\omega h^+}{h^+}-sin\frac{\omega h^-}{h^-})),$$

A and E are as in (1). It is conjectured that R satisfies the resolvent condition of the Kreiss matrix theorem in [10, 11] and hence is power-bounded. The power-boundedness of the operator R, leads to the following theorem on the estimate of the error.

Theorem 1 *Assume that R is power-bounded. Let $\mathbf{u}(x,t)$ be the solution of (1) and $\mathbf{U}^{k+1}(x_j)$ be the solution of (18). Then, for $j = 0, 1, ..., N$ and $k = 0, 1, ..., M-1$,*

$$||\mathbf{u}(x_j, t_{k+1}) - \mathbf{U}^{k+1}(x_j)|| \leq C((N^{-1}\ln N)^2 + M^{-2}),$$

where C is independent of ε, i, j, N and M.

Proof Using Lemmas 8, 15 and 16, it follows that

$$||\mathbf{u}(x_j, t_{k+1}) - \mathbf{U}^{k+1}(x_j)|| \leq C((N^{-1}\ln N)^2 + M^{-2}),$$

as required.

6 Numerical Illustrations

In this section two problems are considered. The Crank-Nicolson method suggested is applied to solve these problems and the results are compared with those obtained by the method suggested in [8], where the first order differential operator in t is discretized using the Backward Euler scheme. To get the order of convergence in the variable t exclusively, a fine Shishkin mesh is considered for x and the resulting problem is solved for various uniform meshes with respect to t. The two mesh algorithm [3] is applied to get the parameter-uniform order of convergence and the error constant.

Next a fine mesh for t is considered, the resulting problem is solved, the $x-$ order of convergence of the method is found. From both the examples the theory that the Crank-Nicolson method doubles the order of convergence with respect to the variable t is well illustrated.

Example 1 Consider the problem

$$\frac{\partial \mathbf{u}}{\partial t} - E\frac{\partial^2 \mathbf{u}}{\partial x^2} + A\mathbf{u} = \mathbf{f} \text{ on } (0,1) \times (0,1], \quad \mathbf{u} = \mathbf{0} \text{ on } \Gamma,$$

where $E = (\varepsilon_1, \varepsilon_2, \varepsilon_3)$, $A = \begin{pmatrix} 6 & -1 & 0 \\ -t & 5(x+1) & -1 \\ -1 & -(1+x^2) & 6+x \end{pmatrix}$, $\mathbf{f} = \begin{pmatrix} 1+e^{x+t} \\ 1+x+t^2 \\ 1+e^t \end{pmatrix}$.

The Crank-Nicolson method is applied to solve the above BVP. For various values of $\varepsilon_1, \varepsilon_2, \varepsilon_3$, the maximum errors, the ε-uniform order of convergence and the

Table 1 Example 1, Crank-Nicolson method, t-convergence

η	Number of mesh points N				
	4	8	16	32	64
0.100E+01	0.666E-01	0.288E-01	0.134E-01	0.532E-02	0.187E-02
0.500E+00	0.634E-01	0.287E-01	0.113E-01	0.376E-02	0.183E-02
0.250E+00	0.637E-01	0.254E-01	0.810E-02	0.370E-02	0.179E-02
0.125E+00	0.602E-01	0.199E-01	0.756E-02	0.361E-02	0.171E-02
0.625E-01	0.529E-01	0.162E-01	0.731E-02	0.346E-02	0.138E-02
0.312E-01	0.453E-01	0.151E-01	0.707E-02	0.282E-02	0.904E-03
D^N	0.666E-01	0.288E-01	0.134E-01	0.532E-02	0.187E-02
p^N	0.121E+01	0.110E+01	0.134E+01	0.151E+01	
C_p^N	0.574E+00	0.532E+00	0.532E+00	0.452E+00	0.341E+00
t-order of convergence = 0.1100358E+01					
The error constant = 0.5736979E+00					

Table 2 Example-1, Euler scheme, t-convergence

η	Number of mesh points N				
	4	8	16	32	64
0.100E+01	0.343E-01	0.230E-01	0.146E-01	0.835E-02	0.449E-02
0.500E+00	0.357E-01	0.238E-01	0.150E-01	0.851E-02	0.455E-02
0.250E+00	0.367E-01	0.243E-01	0.152E-01	0.856E-02	0.458E-02
0.125E+00	0.381E-01	0.248E-01	0.152E-01	0.856E-02	0.456E-02
0.312E-01	0.384E-01	0.248E-01	0.152E-01	0.843E-02	0.447E-02
D^N	0.384E-01	0.248E-01	0.153E-01	0.856E-02	0.458E-02
p^N	0.630E+00	0.700E+00	0.836E+00	0.904E+00	
C_p^N	0.260E+00	0.260E+00	0.248E+00	0.215E+00	0.178E+00
t-order of convergence = 0.6304507E+00					
The error constant = 0.2602176E+00					

ε-uniform error constant are computed using the general methodology from [3]. The variation in all the three parameters is given by considering $\varepsilon_3 = \eta$, $\varepsilon_2 = \frac{\eta}{8}$, $\varepsilon_1 = \frac{\eta}{32}$ where η is varied as shown in the tables. α is taken to be 2.9

Fixing a fine Shishkin mesh with 48 points horizontally, the problem is solved by the Crank-Nicolson method suggested in this paper and the Backward Euler scheme method suggested in [8]. The order of convergence and the error constant are calculated for t and the results are presented in Tables 1 and 2. A fine uniform mesh on t with 32 points is considered and the order of covergence in the variable x is calculated. The results are presented in Table 3. Table 4 presents corresponding results by the method suggested in [8].

Table 3 Example 1, Crank-Nicolson method, x-convergence

η	Number of mesh points N				
	24	48	96	192	384
0.100E+01	0.504E-02	0.132E-02	0.333E-03	0.837E-04	0.209E-04
0.500E+00	0.975E-02	0.259E-02	0.663E-03	0.167E-03	0.418E-04
0.250E+00	0.181E-01	0.503E-02	0.132E-02	0.332E-03	0.834E-04
0.125E+00	0.282E-01	0.973E-02	0.259E-02	0.662E-03	0.167E-03
0.625E-01	0.357E-01	0.181E-01	0.502E-02	0.132E-02	0.332E-03
0.312E-01	0.358E-01	0.282E-01	0.973E-02	0.259E-02	0.662E-03
D^N	0.358E-01	0.282E-01	0.973E-02	0.259E-02	0.662E-03
p^N	0.346E+00	0.154E+01	0.191E+01	0.197E+01	
C_p^N	0.505E+00	0.505E+00	0.221E+00	0.748E-01	0.243E-01
x-order of convergence = $0.3460829E$+00					
The error constant = $0.5048752E$+00					

Table 4 Example 1, Euler scheme, x-convergence

η	Number of mesh points N				
	24	48	96	192	384
0.100E+01	0.458E-02	0.896E-03	0.669E-03	0.416E-03	0.231E-03
0.500E+00	0.951E-02	0.217E-02	0.864E-03	0.583E-03	0.337E-03
0.250E+00	0.179E-01	0.453E-02	0.102E-02	0.776E-03	0.475E-03
0.125E+00	0.281E-01	0.948E-02	0.215E-02	0.971E-03	0.647E-03
0.625E-01	0.357E-01	0.179E-01	0.451E-02	0.112E-02	0.843E-03
0.312E-01	0.359E-01	0.281E-01	0.947E-02	0.214E-02	0.104E-02
D^N	0.359E-01	0.281E-01	0.947E-02	0.214E-02	0.104E-02
p^N	0.352E+00	0.157E+01	0.215E+01	0.104E+01	
C_p^N	0.507E+00	0.507E+00	0.218E+00	0.628E-01	0.389E-01
x-order of convergence = $0.3520847E$+00					
The error constant = $0.5074601E$+00					

Example 2 Consider the problem

$$\frac{\partial \mathbf{u}}{\partial t} - E\frac{\partial^2 \mathbf{u}}{\partial x^2} + A\mathbf{u} = \mathbf{f} \text{ on } (0,1) \times (0,1], \quad \mathbf{u} = \mathbf{0} \text{ on } \Gamma,$$

where $E = (\varepsilon_1, \varepsilon_2, \varepsilon_3)$, $A = \begin{pmatrix} 4(1+x+t) & -t & -x \\ -2(1-t) & 7+((2+t)x) & -(3-x) \\ -1 & -(x+t) & 4(1+\frac{x}{2}+\frac{t}{2}) \end{pmatrix}$,

$\mathbf{f} = (1+x+t^2,\ 1+e^{x+t},\ 2)^T$.

Table 5 Example 2, Crank-Nicolson method, t-convergence

η	Number of mesh points N				
	8	16	32	64	128
0.156E-01	0.312E-01	0.140E-01	0.655E-02	0.260E-02	0.852E-03
0.781E-02	0.293E-01	0.137E-01	0.541E-02	0.173E-02	0.684E-03
0.391E-02	0.295E-01	0.116E-01	0.372E-02	0.139E-02	0.444E-03
0.195E-02	0.265E-01	0.856E-02	0.286E-02	0.914E-03	0.229E-03
0.977E-03	0.215E-01	0.606E-02	0.194E-02	0.486E-03	0.207E-03
0.488E-03	0.210E-01	0.589E-02	0.183E-02	0.487E-03	0.216E-03
D^N	0.312E-01	0.140E-01	0.655E-02	0.260E-02	0.852E-03
p^N	0.116E+01	0.109E+01	0.133E+01	0.161E+01	
C_p^N	0.568E+00	0.542E+00	0.542E+00	0.459E+00	0.320E+00
t-order of convergence = 0.1091209E+01					
The error constant = 0.5679663E+00					

Table 6 Example 2, Euler scheme, t-convergence

η	Number of mesh points N				
	8	16	32	64	128
0.156E-01	0.241E-01	0.152E-01	0.871E-02	0.467E-02	0.242E-02
0.781E-02	0.242E-01	0.153E-01	0.868E-02	0.467E-02	0.243E-02
0.391E-02	0.243E-01	0.153E-01	0.867E-02	0.461E-02	0.238E-02
0.195E-02	0.244E-01	0.153E-01	0.855E-02	0.453E-02	0.234E-02
0.977E-03	0.245E-01	0.153E-01	0.858E-02	0.455E-02	0.235E-02
0.488E-03	0.246E-01	0.154E-01	0.863E-02	0.458E-02	0.236E-02
D^N	0.246E-01	0.154E-01	0.871E-02	0.467E-02	0.243E-02
p^N	0.673E+00	0.824E+00	0.898E+00	0.946E+00	
C_p^N	0.267E+00	0.267E+00	0.241E+00	0.206E+00	0.171E+00
t-order of convergence = 0.6733944E+00					
The error constant = 0.2674448E+00					

The variation in all the three parameters is given by considering $\varepsilon_3 = \eta$, $\varepsilon_2 = \frac{\eta}{4}$, $\varepsilon_1 = \frac{\eta}{16}$ where η is varied as shown in the tables. α is taken to be 0.9.

Fixing a fine Shishkin mesh with 192 points horizontally, the problem is solved by the Crank-Nicolson method suggested in this paper and the Backward Euler scheme method suggested in [8]. A fine uniform mesh on t with 32 points is considered and the order of covergence in the variable x is calculated. Tables 5 and 7 present the results by the Crank Nicolson Method. Tables 6 and 8 present the results by the method presented in [8].

Table 7 Example 2, Crank-Nicolson method, x-convergence

η	Number of mesh points N				
	48	96	192	384	768
0.156E-01	0.945E-02	0.328E-02	0.997E-03	0.285E-03	0.786E-04
0.781E-02	0.157E-01	0.535E-02	0.171E-02	0.574E-03	0.155E-03
0.391E-02	0.216E-01	0.941E-02	0.327E-02	0.997E-03	0.285E-03
0.195E-02	0.238E-01	0.157E-01	0.534E-02	0.171E-02	0.574E-03
0.977E-03	0.239E-02	0.494E-02	0.940E-02	0.327E-02	0.997E-03
0.488E-03	0.238E-02	0.298E-02	0.194E-02	0.174E-02	0.606E-03
D^N	0.238E-01	0.157E-01	0.940E-02	0.327E-02	0.997E-03
p^N	0.599E+00	0.740E+00	0.152E+01	0.171E+01	
C_p^N	0.712E+00	0.712E+00	0.646E+00	0.340E+00	0.157E+00
x-order of convergence = 0.5994747E+00					
The error constant = 0.7119137E+001					

Table 8 Example 2, Euler scheme, x-convergence

η	Number of mesh points N				
	48	96	192	384	768
0.156E-01	0.932E-02	0.323E-02	0.828E-03	0.387E-03	0.222E-03
0.781E-02	0.156E-01	0.531E-02	0.167E-02	0.502E-03	0.304E-03
0.391E-02	0.215E-01	0.929E-02	0.322E-02	0.826E-03	0.406E-03
0.195E-02	0.238E-01	0.156E-01	0.531E-02	0.166E-02	0.522E-03
0.977E-03	0.238E-02	0.489E-02	0.927E-02	0.322E-02	0.825E-03
0.488E-03	0.237E-02	0.291E-02	0.173E-02	0.184E-02	0.824E-03
D^N	0.238E-01	0.156E-01	0.927E-02	0.322E-02	0.825E-03
p^N	0.607E+00	0.751E+00	0.153E+01	0.197E+01	
C_p^N	0.725E+00	0.725E+00	0.656E+00	0.347E+00	0.135E+00
x-order of convergence = 0.6068922E+00					
The error constant = 0.7251142E+00					

Acknowledgments The authors thank the unknown referees for their suggestions and comments which helped us to get the paper in the present form.

References

1. J.J.H. Miller, E. O'Riordan, G.I. Shishkin, *Fitted Numerical Methods for Singular Perturbation Problems* (World Scientific Publishing Co., Singapore, 1996)
2. H.G. Roos , M. Stynes, L. Tobiska, *Numerical Methods for Singularly Perturbed Differential Equations* (Springer, 1996)

3. P.A. Farrell, A. Hegarty, J.J.H. Miller, E. O'Riordan, G.I. Shishkin, *Robust Computational Techniques for Boundary Layers*, eds. by R.J. Knops, K.W. Morton. Applied Mathematics and Mathematical Computation (Chapman & Hall/CRC Press, 2000)
4. J.J.H. Miller, E. O'Riordan, G.I. Shishkin, L.P. Shishkina, Fitted mesh methods for problems with parabolic boundary layers. Math. Proc. R. Irish Acad. **98A**(2), 173–190 (1998)
5. C. Clavero, J.L. Gracia, F.J. Lisbona, Second order uniform approximations for the solution of time dependent singularly perturbed reaction—diffusion systems. Int. J. Numer. Anal. Model. **7**v, 428–443, (2010)
6. T. Linss, *Robust method for a time dependent system of coupled singularly perturbed reaction-diffusion problems*, Preprint MATH-NM-03-2008, TU Dresden, April (2008)
7. J.L. Gracia, F.J. Lisbona, E. O'Riordan, A coupled system of singularly perturbed parabolic reaction-diffusion equations. Adv. Comput. Math. **32**, 43–61 (2010)
8. V. Franklin, M. Paramasivam, S. Valarmathi, J.J.H. Miller, Second order parameter-uniform convergence for a finite difference method for a singularly perturbed linear parabolic system. Int. J. Numer. Anal. Model. **10**(1), 178–202 (2013)
9. M. Paramasivam, S. Valarmathi, J.J.H. Miller, Second order parameter-uniform convergence for a finite difference method for a singularly perturbed linear reaction diffusion system. Math. Commun. **15**(2) (2010)
10. J.J.H. Miller, G. Strang, Matrix theorems for partial differential and difference equations. Mathematica Scandinavica **18**, 113–133 (1966)
11. J.J.H. Miller, On power—bounded operators, and operators satisfying a resolvent condition. Numer. Math. **10**, 389–396 (1967)

A Numerical Method for a System of Singularly Perturbed Differential Equations of Reaction-Diffusion Type with Negative Shift

P. Avudai Selvi and Ramanujam Narasimhan

Abstract A numerical method based on an iterative scheme is proposed for a system of singularly perturbed differential equations of reaction-diffusion type with negative shift term. In this method the solution of the delay problem is obtained as the limit of the solutions to a sequence of the non-delay problems. Then non-delay problems are solved by applying available finite difference scheme and finite element method in the literature. An error estimate in supremum norm is derived. Numerical experiments are carried out.

Keywords System of singularly perturbed problem · Maximum principle · Reaction-diffusion problem · Finite difference scheme · Finite element method · Shishkin mesh · Delay · Negative shift

AMS Mathematics Subject Classification (2010): 34K10 · 34K26 · 34K28

1 Introduction

Delay differential equations appear in various discipline, where they provide good approximations of the observed phenomena. The problems involving these differential equations occur where the future depends not only on the immediate present, but also on the past history of the system under consideration. A delay differential equation is said to be retarded type if the delay argument does not occur in the highest order derivative term. Further if the highest derivative of this delay differential equation is multiplied by a small parameter, then we get singularly perturbed delay differential equations of retarded type. Such type of equations arises frequently

P. Avudai Selvi · R. Narasimhan (✉)
Department of Mathematics, School of Mathematical Sciences,
Bharathidasan University, Tiruchirappalli 620024, Tamil Nadu, India
e-mail: matram2k3@gmail.com

P. Avudai Selvi
e-mail: avudaiselvi12@gmail.com

V. Sigamani et al. (eds.), *Differential Equations and Numerical Analysis*,
Springer Proceedings in Mathematics & Statistics 172,
DOI 10.1007/978-81-322-3598-9_6

in the mathematical modeling of various practical phenomena, for example, in the modeling of the human pupil-light reflex [1], the study of bistable devices [2] and variational problems in control theory [3], etc.

In general the standard numerical methods for solving singularly perturbation problems are sometimes not stable and also they do not give desired results when the perturbation parameter is very small. Hence many people developed suitable numerical methods to solve singularly perturbed differential equations such that these methods provide good accuracy which do not depend on the small parameter. In the literature these methods are known as parameter-uniform numerical methods.

In the recent years there has been a growing interest in the area of numerical methods to SPDDEs. In fact, Erdogan [4] proposed an exponentially fitted operator method for singularly perturbed first order equations. Kadalbajoo and Sharma [5–7] and Mohapatra and Natesan [8] proposed some methods for equations of reaction-diffusion type with small delays either in function or its derivative.

Subburayan and Ramanujam [9–14] suggested two methods namely initial value technique and asymptotic numerical method for equations of reaction-diffusion type as well as convection-diffusion type. Nicaise and Xenophontos [15], developed a hp-version finite element method for reaction-diffusion type DEs. In [16], the author proposed a discontinuous Galerkin finite element method for reaction-diffusion type DEs and established robust convergence of the method in the energy norm.

We, in this paper, using the iterative procedure given in [17], finite difference scheme and finite element method available in the literature for system of singularly perturbed differential equations without delay, propose a numerical method to find a numerical solution for the following problem:

Find $\overline{u} = (u_1,\ u_2),\ u_1,\ u_2 \in U := C^2(\Omega) \cap C(\overline{\Omega})$ such that

$$\begin{cases} -\varepsilon u_1''(x) + \sum_{k=1}^{2} a_{1k}(x)u_k(x) + \sum_{k=1}^{2} b_{1k}(x)u_k(x-1) = f_1(x), & x \in \Omega, \\ -\varepsilon u_2''(x) + \sum_{k=1}^{2} a_{2k}(x)u_k(x) + \sum_{k=1}^{2} b_{2k}(x)u_k(x-1) = f_2(x), & x \in \Omega, \\ u_1(x) = \phi_1(x),\ x \in [-1,0],\ u_1(2) = l_1, \\ u_2(x) = \phi_2(x),\ x \in [-1,0],\ u_2(2) = l_2, \end{cases} \tag{1}$$

where $0 < \varepsilon \le 1, a_{11}(x) > 0, a_{22}(x) > 0, a_{12}(x) \le 0, a_{21}(x) \le 0, a_{i1}(x) + a_{i2}(x) \ge \alpha_i \ge \alpha > 0,\ i = 1, 2, b_{ij}(x) \le 0,\ i = 1, 2,\ j = 1, 2,\ 0 > b_{i1}(x) + b_{i2}(x) \ge -\beta_i \ge -\beta,\ i = 1, 2,\ \alpha - \beta > 0$, the functions $a_{ik},\ b_{ik},\ f_i \in C^4(\overline{\Omega}),\ i = 1, 2,\ k = 1, 2,$ $\Omega = (0, 2), \overline{\Omega} = [0, 2], \Omega^- = (0, 1), \Omega^+ = (1, 2)$ and $\phi_i \in C^4([-1, 0]),\ i = 1, 2.$

The present paper is organized as follows. Section 2 presents the maximum principle and the stability result. The proposed iterative procedure is explained in Sect. 3. Mesh selection strategy is discussed in Sect. 4. A first order finite difference scheme is presented in Sect. 5, where as the Sect. 6 deals with the finite element method. Numerical experiments are carried out in Sect. 7. The paper concludes with a discussion.

Throughout our analysis C is a generic positive constant that is independent of parameter ε and number of mesh points N. In this paper the following supremum norm is used:

$$\|u\|_D = \sup_{x \in D} \mid u(x) \mid .$$

In case of vectors $\overline{u} = (u_1, u_2)$, we define

$$\|\overline{u}\| = \max\{\|u_1\|,\ \|u_2\|\}.$$

2 Maximum Principle and Stability Result

Consider the following problem.

Find $\overline{u} = (u_1,\ u_2),\ u_1,\ u_2 \in U^* := C^2(\Omega^- \cup \Omega^+) \cap C(\overline{\Omega})$ such that

$$\begin{cases} P_1\overline{u}(x): \ = \begin{cases} -\varepsilon u_1''(x) + \sum_{k=1}^{2} a_{1k}(x)u_k(x) = f_1(x) - \sum_{k=1}^{2} b_{1k}(x)\phi_k(x-1), & x \in \Omega^-, \\ -\varepsilon u_1''(x) + \sum_{k=1}^{2} a_{1k}(x)u_k(x) + \sum_{k=1}^{2} b_{1k}(x)u_k(x-1) = f_1(x), & x \in \Omega^+, \end{cases} \\ P_2\overline{u}(x): \ = \begin{cases} -\varepsilon u_2''(x) + \sum_{k=1}^{2} a_{2k}(x)u_k(x) = f_2(x) - \sum_{k=1}^{2} b_{2k}(x)\phi_k(x-1), & x \in \Omega^-, \\ -\varepsilon u_2''(x) + \sum_{k=1}^{2} a_{2k}(x)u_k(x) + \sum_{k=1}^{2} b_{2k}(x)u_k(x-1) = f_2(x), & x \in \Omega^+, \end{cases} \\ u_1(0) = \phi_1(0),\ u_1(1-) = u_1(1+),\ u_1'(1-) = u_1'(1+),\ u_1(2) = l_1, \\ u_2(0) = \phi_2(0),\ u_2(1-) = u_2(1+),\ u_2'(1-) = u_2'(1+),\ u_2(2) = l_2. \end{cases} \tag{2}$$

The differential-difference operator $P = (P_1, P_2)$ satisfies the following maximum principle.

Theorem 1 (Maximum Principle) *Let $\overline{w} = (w_1,\ w_2),\ w_1,\ w_2 \in U^*$ be any function satisfying $w_i(0) \geq 0,\ w_i(2) \geq 0,\ P_i\overline{w}(x) \geq 0,\ \forall x \in \Omega^- \cup \Omega^+$ and $w_i'(1+) - w_i'(1-) = [w_i'](1) \leq 0,\ i = 1, 2$. Then $w_i(x) \geq 0,\ \forall x \in \overline{\Omega},\ i = 1, 2$.*

Proof Using the method of proof given in Theorem 3.1 of [12], one can prove the present theorem. □

Note The above theorem was proved using the conditions $\alpha - \beta > 0$ [12].
The following stability result can be proved by using the above maximum principle.

Corollary 1 *Let $\overline{w} = (w_1,\ w_2),\ w_1,\ w_2 \in U^*$. Then*

$$\mid w_j(x) \mid \leq C \max\left\{\max_{i=1,2}\{\mid w_i(0) \mid\},\ \max_{i=1,2}\{\mid w_i(2) \mid\},\ \max_{i=1,2}\{\| P_i\overline{w} \|_{\Omega^- \cup \Omega^+}\}\right\}, \quad \forall\, x \in \overline{\Omega},\ j = 1, 2, \tag{3}$$

where $C = 8 \max\left\{1,\ \frac{1}{\alpha},\ \frac{1}{5(\alpha-\beta)}\right\}$.

Proof Using the method of proof given in Corollary 3.2 of [12], one can prove the present corollary. □

3 Iterative Method

3.1 Procedure

Using the iterative procedure suggested in [17], we define the following iterative method for the above boundary value problem (2).

Let $\overline{u}^{(0)} = (u_1^{(0)},\ u_2^{(0)})$ such that

$$\begin{cases} u_1^{(0)}(x) = \phi_1(x), & x \in [-1,0], \\ u_1^{(0)}(x) = \phi_1(0), & x \in [0,2], \end{cases} \quad \begin{cases} u_2^{(0)}(x) = \phi_2(x), & x \in [-1,0], \\ u_2^{(0)}(x) = \phi_2(0), & x \in [0,2], \end{cases} \tag{4}$$

and

$$\begin{cases} \overline{u}^{(n)} = (u_1^{(n)},\ u_2^{(n)}),\ \ u_1^{(n)},\ u_2^{(n)} \in U^* \text{ such that} \\ u_1^{(n)}(x) = \phi_1(x), \quad x \in [-1,0], \\ u_2^{(n)}(x) = \phi_2(x), \quad x \in [-1,0], \\ -\varepsilon u_1^{(n)''}(x) + \sum_{k=1}^{2} a_{1k}(x) u_k^{(n)}(x) = \begin{cases} f_1(x) - \sum_{k=1}^{2} b_{1k}(x)\phi_k(x-1), & x \in \Omega^-, \\ f_1(x) - \sum_{k=1}^{2} b_{1k}(x) u_k^{(n-1)}(x-1), & x \in \Omega^+, \end{cases} \\ -\varepsilon u_2^{(n)''}(x) + \sum_{k=1}^{2} a_{2k}(x) u_k^{(n)}(x) = \begin{cases} f_2(x) - \sum_{k=1}^{2} b_{2k}(x)\phi_k(x-1), & x \in \Omega^-, \\ f_2(x) - \sum_{k=1}^{2} b_{2k}(x) u_k^{(n-1)}(x-1), & x \in \Omega^+, \end{cases} \\ u_1^{(n)}(0) = \phi_1(0), \quad u_1^{(n)}(1-) = u_1^{(n)}(1+), \quad u_1^{(n)'}(1-) = u_1^{(n)'}(1+), \quad u_1^{(n)}(2) = l_1, \\ u_2^{(n)}(0) = \phi_2(0), \quad u_2^{(n)}(1-) = u_2^{(n)}(1+), \quad u_2^{(n)'}(1-) = u_2^{(n)'}(1+), \quad u_2^{(n)}(2) = l_2, \end{cases} \tag{5}$$

for $n = 1,\ 2,\ \cdots$.

3.2 Convergence Analysis

To prove that the sequence defined in the previous section converges uniformly to the solution of the problem (2) the following result is used.

Theorem 2 (Maximum Principle) *Let* $\overline{w} = (w_1,\ w_2),\ w_1,\ w_2 \in U^*$ *such that* $w_i(0) \geq 0,\ w_i(2) \geq 0,\ P_i^*\ \overline{w}(x) \geq 0,\ \forall x \in \Omega^- \cup \Omega^+$ *and* $w_i'(1+) - w_i'(1-) = [w_i'](1) \leq 0,\ i = 1, 2,$ *where* $P_i^*\ \overline{w}(x) = -\varepsilon w_i''(x) + \sum_{k=1}^{2} a_{ik}(x) w_k(x),\ i = 1, 2.$ *Then* $w_i(x) \geq 0,\ \forall x \in \overline{\Omega},\ i = 1, 2.$

Proof For proof refer (Theorem 2.2, [19]). □

Theorem 3 (Uniform Convergence) *The sequence* $\{\overline{u}^{(n)}\}$ *defined by (4)–(5) converges uniformly to the solution* $\overline{u}$ *of the problem (2).*

Proof From Corollary 1, we have

$$|u_j(x)| \le C \max\left\{\max_{i=1,2}\{|u_i(0)|\},\ \max_{i=1,2}\{|u_i(2)|\},\ ||\overline{f}|| + \beta||\overline{\phi}||\}\right\} = M' \text{ (say)}, \quad \forall\, x \in \overline{\Omega},\ j = 1, 2.$$

Let $\overline{z}^{(n)} = (z_1^{(n)}, z_2^{(n)})$, $z_i^{(n)}(x) = u_i^{(n)}(x) - u_i(x),\ n = 0, 1, 2, \cdots,\ i = 1, 2$. Then, on $\overline{\Omega}$, we have

$$\begin{aligned} |z_i^{(0)}(x)| = |u_i^{(0)}(x) - u_i(x)| &= |\phi_i(0) - u_i(x)|, \\ &\le |\phi_i(0)| + |u_i(x)|, \\ &\le |\phi_i(0)| + M' = M \text{ (say)}, \quad i = 1, 2. \end{aligned}$$

We have

$$\begin{cases} P_1^* \overline{z}^{(n)}(x) = \begin{cases} 0, & x \in \Omega^-, \\ -\sum_{k=1}^2 b_{1k}(x) z_k^{(n-1)}(x-1), & x \in \Omega^+, \end{cases} \\ P_2^* \overline{z}^{(n)}(x) = \begin{cases} 0, & x \in \Omega^-, \\ -\sum_{k=1}^2 b_{2k}(x) z_k^{(n-1)}(x-1), & x \in \Omega^+, \end{cases} \\ z_1^{(n)}(0) = 0, \quad z_1^{(n)}(2) = 0, \\ z_2^{(n)}(0) = 0, \quad z_2^{(n)}(2) = 0, \quad n = 1,\ 2,\ \cdots. \end{cases}$$

Define, $\overline{\eta}_n = (\eta_n, \eta_n)$, where $\eta_n = \left(\frac{\beta}{\alpha}\right)^n M,\quad n = 0, 1, 2, \cdots$. Since $\alpha - \beta > 0$, then $\eta_n \to 0$ as $n \to \infty$. By the method of induction we now prove that

$$||z_i^{(n)}||_{\overline{\Omega}} \le \eta_n,\ i = 1, 2, \quad n = 0, 1, 2, \cdots.$$

Clearly, $|z_i^{(0)}(x)| \le M = \eta_0,\ \forall x \in \overline{\Omega},\ i = 1, 2$. Therefore, $||z_i^{(0)}||_{\overline{\Omega}} \le \eta_0,\ i = 1, 2$.

We now prove that $||z_i^{(1)}||_{\overline{\Omega}} \le \eta_1,\ i = 1, 2$. We have

$$\begin{aligned} P_i^* \overline{\eta}_1 &= \textstyle\sum_{k=1}^2 a_{ik}(x)\eta_1 \ge \alpha\eta_1 = \beta M \ge 0 = P_i^* \overline{z}^{(1)}(x), \quad x \in \Omega^-, \\ P_i^* \overline{\eta}_1 &= \textstyle\sum_{k=1}^2 a_{ik}(x)\eta_1 \ge \alpha\eta_1 = \beta M \ge -\sum_{k=1}^2 b_{ik}(x) M \\ &\ge -\textstyle\sum_{k=1}^2 b_{ik}(x) z_k^{(0)}(x-1) = P_i^* \overline{z}^{(1)}(x), \qquad x \in \Omega^+,\quad i = 1, 2. \end{aligned}$$

Therefore, $P_i^* \overline{z}^{(1)}(x) \le P_1^* \overline{\eta}_1, \forall x \in \Omega^- \cup \Omega^+, i = 1, 2$. Also $z_i^{(1)}(0) = 0 \le \eta_1$, $z_i^{(1)}(2) = 0 \le \eta_1, [z_i^{(1)'}](1) = 0, i = 1, 2$. Then, by Theorem 2, we have $z_i^{(1)}(x) \le \eta_1$, $\forall x \in \overline{\Omega}, i = 1, 2$. Similarly we can show that $-\eta_1 \le z_i^{(1)}(x),\ \forall x \in \overline{\Omega}, i = 1, 2$. Therefore,

$$||z_i^{(1)}||_{\overline{\Omega}} \le \eta_1, \quad i = 1, 2. \tag{6}$$

Now assume that, $||z_i^{(n)}||_{\overline{\Omega}} \leq \eta_n,\ i = 1, 2$, for some $n > 1$. We have

$$\begin{aligned}
P_i^* \,\overline{\eta}_{n+1} &= \textstyle\sum_{k=1}^{2} a_{ik}(x)\eta_{n+1} \geq \alpha\eta_{n+1} = \beta\left(\frac{\beta}{\alpha}\right)^n M \geq 0 = P_i^* \,\overline{z}^{(n+1)}(x), \quad x \in \Omega^-, \\
P_i^* \,\overline{\eta}_{n+1} &= \textstyle\sum_{k=1}^{2} a_{ik}(x)\eta_{n+1} \geq \alpha\eta_{n+1} = \beta\left(\frac{\beta}{\alpha}\right)^n M \geq -\textstyle\sum_{k=1}^{2} b_{ik}(x)\eta_n \\
&\geq -\textstyle\sum_{k=1}^{2} b_{ik}(x) z_k^{(n)}(x-1) = P_i^* \,\overline{z}^{(n+1)}(x), \quad x \in \Omega^+, \quad i = 1, 2.
\end{aligned}$$

Therefore, $P_i^* \,\overline{z}^{(n+1)}(x) \leq P_i^* \,\overline{\eta}_{n+1},\ \forall x \in \Omega^- \cup \Omega^+, i = 1, 2$. Also $z_i^{(n+1)}(0) = 0 \leq \eta_{n+1},\ z_i^{(n+1)}(2) = 0 \leq \eta_{n+1},\ [z_i^{(n+1)'}](1) = 0, i = 1, 2$. Then, by Theorem 2, we have $z_i^{(n+1)}(x) \leq \eta_{n+1},\ \forall x \in \overline{\Omega},\quad i = 1, 2$. Similarly we can show that $-\eta_{n+1} \leq z_i^{(n+1)}(x),\ \forall x \in \overline{\Omega}, i = 1, 2$. Therefore,

$$||z_i^{(n+1)}||_{\overline{\Omega}} \leq \eta_{n+1}, \quad i = 1, 2. \tag{7}$$

Hence the sequence $\{\overline{u}^{(n)}\}$ converges uniformly to the solution of the problem (2).□

Note: Theorem 1 was proved under the assumption that $\alpha - \beta > 0$ [12]. Further it may be observed that the same assumption is used to prove the uniform convergence. That is, no extra assumption is imposed on the coefficients.

4 Layer Adapted Meshes

In this section we present two types of meshes namely Shishkin mesh and Bakhvalov-Shishkin mesh (BS mesh).

4.1 *Shishkin Mesh*

The present paper is an extension of the work carried out in [18] from single to system. In [18] authors mentioned that the differential equation exhibits boundary layers at $x = 0, x = 2$ and interior layers (left and right) at $x = 1$. The same conclusion can be made for the above problem (1) by deriving suitable estimates for the solution. Therefore we divide the interval $[0, 2]$ into six subintervals, namely $\Omega_1 = [0, \tau]$, $\Omega_2 = [\tau, 1-\tau]$, $\Omega_3 = [1-\tau, 1]$, $\Omega_4 = [1, 1+\tau]$, $\Omega_5 = [1+\tau, 2-\tau]$ and $\Omega_6 = [2-\tau, 2]$, where $\tau = \min\left\{\frac{1}{4}, \frac{2\sqrt{\varepsilon}\ln(N)}{\sqrt{\alpha}}\right\}$. On Ω_1, Ω_3, Ω_4 and Ω_6 a uniform mesh with $\frac{N}{8}$ mesh intervals is placed, while on Ω_2 and Ω_5 uniform mesh with $\frac{N}{4}$ mesh intervals is placed.

The Shishkin mesh $\overline{\Omega}^N = \{x_0,\ x_1,\ \cdots,\ x_N\}$ is defined by

$$\begin{cases} x_0 = 0, \\ x_i = x_0 + ih, & 1 \le i \le \frac{N}{8}, \\ x_{i+\frac{N}{8}} = x_{\frac{N}{8}} + iH, & 1 \le i \le \frac{N}{4}, \\ x_{i+\frac{3N}{8}} = x_{\frac{3N}{8}} + ih, & 1 \le i \le \frac{N}{8}, \\ x_{i+\frac{N}{2}} = x_{\frac{N}{2}} + ih, & 1 \le i \le \frac{N}{8}, \\ x_{i+\frac{5N}{8}} = x_{\frac{5N}{8}} + iH, & 1 \le i \le \frac{N}{4}, \\ x_{i+\frac{7N}{8}} = x_{\frac{7N}{8}} + ih, & 1 \le i \le \frac{N}{8}. \end{cases}$$

where $h = 8N^{-1}\tau$ and $H = 4N^{-1}(1 - 2\tau)$.

4.2 BS Mesh

The mesh points of these meshes are given by [20]

$$x_i = \begin{cases} 2\sqrt{\frac{\varepsilon}{\alpha}}\phi_1(t_i), & i = 0, \cdots, N/8 \\ \tau + \frac{4}{N}(1-2\tau)\big(i - \frac{N}{8}\big), & i = N/8+1, \cdots, 3N/8 \\ 1 - 2\sqrt{\frac{\varepsilon}{\alpha}}\phi_2(t_i), & i = 3N/8+1, \cdots, N/2 \\ 1 + 2\sqrt{\frac{\varepsilon}{\alpha}}\phi_3(t_i), & i = N/2+1, \cdots, 5N/8 \\ 1 + \tau + \frac{4}{N}(1-2\tau)\big(i - \frac{5N}{8}\big), & i = 5N/8+1, \cdots, 7N/8 \\ 2 - 2\sqrt{\frac{\varepsilon}{\alpha}}\phi_4(t_i), & i = 7N/8+1, \cdots, N \end{cases}$$

where $t_i = i/N$, $\phi_i = -\ln\psi_i$, $i = 1, 2, 3, 4$, $\psi_1(t) = 1 - 8(1 - N^{-1})t$, $\psi_2(t) = 1 - 4(1 - N^{-1})(1 - 2t)$, $\psi_3(t) = 1 - 4(1 - N^{-1})(2t - 1)$ and $\psi_4(t) = 1 - 8(1 - N^{-1})(1 - t)$.

5 First Order Finite Difference Scheme

In the present section a finite difference scheme is presented for problems (5). Further we derive an error estimate for the numerical method suggested in this paper.

5.1 Scheme

Using the finite difference scheme discussed in [19] on the Shishkin mesh $\overline{\Omega}^N = \{x_0,\ x_1,\ \cdots,\ x_N\}$, we now define the following finite difference scheme for the sequence of the problems (5).

Let $\overline{U}^{[0]} = (U_1^{[0]}, U_2^{[0]}),\ U_i^{[0]} = \big(u_i^{(0)}(x_0),\ u_i^{(0)}(x_1),\ \cdots,\ u_i^{(0)}(x_N)\big), i = 1, 2.$
Find $\overline{U}^{[n]} = (U_1^{[n]}, U_2^{[n]}),\ U_i^{[n]} = \big(U_{i,1}^{[n]},\ U_{i,2}^{[n]}, \cdots, U_{i,N}^{[n]}\big), i = 1, 2$, such that

$$\begin{cases} P_1^{*N}\overline{U}_j^{[n]} \equiv -\varepsilon\delta^2 U_{1,j}^{[n]} + \sum_{k=1}^{2} a_{1k,j}U_{k,j}^{[n]} = \begin{cases} f_{1,j} - \sum_{k=1}^{2} b_{1k,j}\phi_k(x_j - 1), & j = 1, \cdots, \frac{N}{2} - 1, \\ f_{1,j} - \sum_{k=1}^{2} b_{1k,j}U_{k,j-\frac{N}{2}}^{[n-1]}, & j = \frac{N}{2} + 1, \cdots, N-1, \end{cases} \\ P_2^{*N}\overline{U}_j^{[n]} \equiv -\varepsilon\delta^2 U_{2,j}^{[n]} + \sum_{k=1}^{2} a_{2k,j}U_{k,j}^{[n]} = \begin{cases} f_{2,j} - \sum_{k=1}^{2} b_{2k,j}\phi_k(x_j - 1), & j = 1, \cdots, \frac{N}{2} - 1, \\ f_{2,j} - \sum_{k=1}^{2} b_{2k,j}U_{k,j-\frac{N}{2}}^{[n-1]}, & j = \frac{N}{2} + 1, \cdots, N-1, \end{cases} \\ U_{1,0}^{[n]} = \phi_1(0), \quad D^+U_{1,\frac{N}{2}}^{[n]} = D^-U_{1,\frac{N}{2}}^{[n]}, \quad U_{1,N}^{[n]} = l_1, \\ U_{2,0}^{[n]} = \phi_2(0), \quad D^+U_{2,\frac{N}{2}}^{[n]} = D^-U_{2,\frac{N}{2}}^{[n]}, \quad U_{2,N}^{[n]} = l_2, \quad n = 1,\ 2,\ \cdots. \end{cases} \tag{8}$$

Here

$$\overline{U}_j^{[n]} = (U_{1,j}^{[n]}, U_{2,j}^{[n]}), \quad U_{1,j}^{[n]} = U_1^{[n]}(x_j), \quad U_{2,j}^{[n]} = U_2^{[n]}(x_j),$$

$$\delta^2 U_{i,j}^{[n]} = \frac{1}{x_{j+1} - x_{j-1}}\left(\frac{U_{i,j+1}^{[n]} - U_{i,j}^{[n]}}{x_{j+1} - x_j} - \frac{U_{i,j}^{[n]} - U_{i,j-1}^{[n]}}{x_j - x_{j-1}}\right),$$

$$D^+U_{i,\frac{N}{2}}^{[n]} = \frac{U_{i,\frac{N}{2}+1}^{[n]} - U_{i,\frac{N}{2}}^{[n]}}{x_{\frac{N}{2}+1} - x_{\frac{N}{2}}} \quad \text{and} \quad D^-U_{i,\frac{N}{2}}^{[n]} = \frac{U_{i,\frac{N}{2}}^{[n]} - U_{i,\frac{N}{2}-1}^{[n]}}{x_{\frac{N}{2}} - x_{\frac{N}{2}-1}},$$

$$a_{ik,j} = a_{ik}(x_j), \quad b_{ik,j} = b_{ik}(x_j) \quad f_{k,j} = f_k(x_j).$$

5.2 Error Estimate

We now derive an error estimate for the above scheme.

Theorem 4 *Let $\overline{u}$ and $\overline{U}^{[n]}$ be the solutions of the problems (2) and (8) respectively. Then we have*

$$\|u_i - U_i^{[n]}\|_{\overline{\Omega}^N} \leq CN^{-1}\ln N, \quad i = 1, 2, \quad \textit{provided that} \quad n \geq \frac{\ln(N^{-1}\ln N)}{\ln\gamma}, \quad \gamma = \frac{\beta}{\alpha}.$$

Proof We have

$$\begin{cases} -\varepsilon u_i^{(1)\prime\prime}(x) + \sum_{k=1}^{2} a_{ik}(x)u_k^{(1)}(x) = \begin{cases} f_i(x) - \sum_{k=1}^{2} b_{ik}(x)\phi_k(x-1), & x \in \Omega^-, \\ f_i(x) - \sum_{k=1}^{2} b_{ik}(x)u_k^{(0)}(x-1), & x \in \Omega^+, \end{cases} \\ u_i^{(1)}(0) = \phi_i(0), \quad u_i^{(1)}(2) = l_i, \quad i = 1, 2. \end{cases} \tag{9}$$

Recall that $U_i^{[0]} = \big(u_i^{(0)}(x_0),\ u_i^{(0)}(x_1),\ \cdots,\ u_i^{(0)}(x_N)\big), i = 1, 2$. From (8) we have

$$\begin{cases} -\varepsilon\delta^2 U_{i,j}^{[1]} + \sum_{k=1}^{2} a_{ik,j} U_{k,j}^{[1]} = \begin{cases} f_{i,j} - \sum_{k=1}^{2} b_{ik,j}\phi_k(x_j - 1), & j = 1, \cdots, \frac{N}{2} - 1, \\ f_{i,j} - \sum_{k=1}^{2} b_{ik,j} u_k^{(0)}(x_j - 1), & j = \frac{N}{2} + 1, \cdots, N - 1, \end{cases} \\ U_{i,0}^{[1]} = \phi_i(0), \quad D^+ U_{i,\frac{N}{2}}^{[1]} = D^- U_{i,\frac{N}{2}}^{[1]}, \quad U_{i,N}^{[1]} = l_i, \quad i = 1, 2. \end{cases} \tag{10}$$

Applying the error estimate result from [19] to the problems (9)–(10) we have

$$|u_i^{(1)}(x_j) - U_{i,j}^{[1]}| \le CN^{-1}\ln N, \quad \text{for } j = 0, 1, \cdots, N \quad i = 1, 2. \tag{11}$$

From (6) and (11) we have

$$\begin{aligned} |u_i(x_j) - U_{i,j}^{[1]}| &\le |u_i(x_j) - u_i^{(1)}(x_j)| + |u_i^{(1)}(x_j) - U_{i,j}^{[1]}| \\ &\le C(\tfrac{\beta}{\alpha}) + CN^{-1}\ln N, \qquad \text{for } j = 0, 1, \cdots, N, \quad i = 1, 2. \end{aligned}$$

Hence

$$\|u_i - U_i^{[1]}\|_{\overline{\Omega}^N} \le C\left(\frac{\beta}{\alpha}\right) + CN^{-1}\ln N, \quad i = 1, 2.$$

Now consider

$$\begin{cases} -\varepsilon u_i^{(2)''}(x) + \sum_{k=1}^{2} a_{ik}(x) u_k^{(2)}(x) = \begin{cases} f_i(x) - \sum_{k=1}^{2} b_{ik}(x)\phi_k(x - 1), & x \in \Omega^-, \\ f_i(x) - \sum_{k=1}^{2} b_{ik}(x) u_k^{(1)}(x - 1), & x \in \Omega^+, \end{cases} \\ u_i^{(2)}(0) = \phi_i(0), \quad u_i^{(2)}(2) = l_i, \quad i = 1, 2. \end{cases} \tag{12}$$

Discretize (12) we get

$$\begin{cases} -\varepsilon\delta^2 U{*}_{i,j}^{[2]} + \sum_{k=1}^{2} a_{ik,j} U{*}_{k,j}^{[2]} = \begin{cases} f_{i,j} - \sum_{k=1}^{2} b_{ik,j}\phi_k(x_j - 1), & j = 1, \cdots, \frac{N}{2} - 1, \\ f_{i,j} - \sum_{k=1}^{2} b_{ik,j} u_k^{(1)}(x_j - 1), & j = \frac{N}{2} + 1, \cdots, N - 1, \end{cases} \\ U{*}_{i,0}^{[2]} = \phi_i(0), \quad D^+ U{*}_{i,\frac{N}{2}}^{[2]} = D^- U{*}_{i,\frac{N}{2}}^{[2]}, \quad U{*}_{i,N}^{[2]} = l_i, \quad i = 1, 2. \end{cases} \tag{13}$$

Applying the error estimate result from [19] to the problems (12)–(13) we have

$$|u_i^{(2)}(x_j) - U{*}_{i,j}^{[2]}| \le CN^{-1}\ln N, \quad \text{for } j = 0, 1, \cdots, N \quad i = 1, 2. \tag{14}$$

Replacing $u_k^{(1)}(x_j - 1)$ by its numerical solution we get

$$\begin{cases} -\varepsilon\delta^2 U_{i,j}^{[2]} + \sum_{k=1}^{2} a_{ik,j} U_{k,j}^{[2]} = \begin{cases} f_{i,j} - \sum_{k=1}^{2} b_{ik,j}\phi_k(x_j - 1), & j = 1, \cdots, \frac{N}{2} - 1, \\ f_{i,j} - \sum_{k=1}^{2} b_{ik,j} U_{k,j-\frac{N}{2}}^{[1]}, & j = \frac{N}{2} + 1, \cdots, N - 1, \end{cases} \\ U_{i,0}^{[2]} = \phi_i(0), \quad D^+ U_{i,\frac{N}{2}}^{[2]} = D^- U_{i,\frac{N}{2}}^{[2]}, \quad U_{i,N}^{[2]} = l_i, \quad i = 1, 2. \end{cases} \tag{15}$$

Applying the discrete stability result [19] for $\overline{Z}^{[2]} = (Z_1^{[2]}, Z_2^{[2]}), Z_{i,j}^{[2]} = U{*}_{i,j}^{[2]} - U_{i,j}^{[2]}$ where $\overline{Z}^{[2]}$ satisfies the following discrete problem

$$\begin{cases} P_i^{*N}\,\overline{Z}_j^{[2]} = \begin{cases} 0, & j = 1, \cdots, \frac{N}{2} - 1, \\ -\sum_{k=1}^{2} b_{ik,j}\left(u_k^{(1)}(x_j - 1) - U_{k,j-\frac{N}{2}}^{[1]}\right), & j = \frac{N}{2} + 1, \cdots, N - 1, \end{cases} \\ Z_{i,0}^{[2]} = 0, \quad D^+ Z_{i,\frac{N}{2}}^{[2]} = D^- Z_{i,\frac{N}{2}}^{[2]}, \quad Z_{i,N}^{[2]} = 0, \ i = 1, 2, \end{cases}$$

we get

$$|U^{*[2]}_{i,j} - U_{i,j}^{[2]}| \leq \frac{\beta}{\alpha} CN^{-1} \ln N < CN^{-1} \ln N, \quad \text{for } j = 0, 1, \cdots, N, \ i = 1, 2. \tag{16}$$

From (14) and (16) we have

$$\begin{aligned} |u_i^{(2)}(x_j) - U_{i,j}^{[2]}| &\leq |u_i^{(2)}(x_j) - U^{*[2]}_{i,j}| + |U^{*[2]}_{i,j} - U_{i,j}^{[2]}| \\ &\leq CN^{-1} \ln N + CN^{-1} \ln N \\ &\leq CN^{-1} \ln N, \qquad \text{for } j = 0, 1, \cdots, N, \ i = 1, 2. \end{aligned}$$

Thus

$$\begin{aligned} |u_i(x_j) - U_{i,j}^{[2]}| &\leq |u_i(x_j) - u_i^{(2)}(x_j)| + |u_i^{(2)}(x_j) - U_{i,j}^{[2]}| \\ &\leq C(\tfrac{\beta}{\alpha})^2 + CN^{-1} \ln N, \qquad \text{for } j = 0, 1, \cdots, N, \ i = 1, 2. \end{aligned}$$

Hence

$$\|u_i - U_i^{[2]}\|_{\overline{\Omega}^N} \leq C\left(\frac{\beta}{\alpha}\right)^2 + CN^{-1} \ln N, \quad i = 1, 2.$$

Continuing this process one can prove that

$$\|u_i - U_i^{[n]}\|_{\overline{\Omega}^N} \leq C\left(\frac{\beta}{\alpha}\right)^n + CN^{-1} \ln N, \quad i = 1, 2, \quad \text{for } n > 2.$$

Since $\frac{\ln(N^{-1} \ln N)}{\ln \gamma} \leq n$ and $\gamma = \frac{\beta}{\alpha}$, we have $(\frac{\beta}{\alpha})^n \leq N^{-1} \ln N$. Finally we have

$$\|u_i - U_i^{[n]}\|_{\overline{\Omega}^N} \leq CN^{-1} \ln N, \quad i = 1, 2.$$

Hence the proof of the theorem. □

Note: The scheme discussed in [19] is for the system of differential equations with discontinuous source terms. It may be noted that though the systems of differential equations in (5) has continuous source terms but the source terms are not differentiable in general. Hence one can use the scheme given in [19] to the problems (5).

6 Finite Element Method

6.1 Scheme

Using the finite element method presented in [20] we get the following system of equations for the problems (5).

$$\begin{cases} P_i^{*N}\overline{U}_j^{[n]} \equiv -\varepsilon\left(\frac{U_{i,j+1}^{[n]}-U_{i,j}^{[n]}}{h_{j+1}} - \frac{U_{i,j}^{[n]}-U_{i,j-1}^{[n]}}{h_j}\right) + \overline{h}_j \sum_{k=1}^{2} a_{ik,j} U_{k,j}^{[n]} = F_{i,j} \\ U_{i,0}^{[n]} = U_{i,N}^{[n]} = 0, \quad i = 1, 2, \end{cases} \tag{17}$$

where

$$F_{i,j} = \begin{cases} \overline{h}_j(f_{i,j} - \sum_{k=1}^{2} b_{ik,j}\phi_k(x_j - 1)), & j = 1, \cdots, \frac{N}{2} - 1, \\ \frac{\overline{h}_j}{2}(f_{i,j-1} - \sum_{k=1}^{2} b_{ik,j-1}\phi_k(x_{j-1} - 1) \\ \qquad + f_{i,j+1} - \sum_{k=1}^{2} b_{ik,j+1}U_{k,1}^{[n-1]}), & j = \frac{N}{2}, \\ \overline{h}_j(f_{i,j} - \sum_{k=1}^{2} b_{ik,j}U_{k,j-\frac{N}{2}}^{[n-1]}), & j = \frac{N}{2} + 1, \cdots, N - 1, \end{cases}$$

for $n = 1,\ 2,\ \cdots$.

The above difference operator $P^{*N} = (P_1^{*N}, P_2^{*N})$ satisfies the following discrete maximum principle.

Theorem 5 (Discrete Maximum Principle) *Let $\overline{W} = (W_1,\ W_2)$ be any mesh function satisfying $P_i^{*N}\overline{W}_j \geq 0$, $\forall\ j = 1, 2, \cdots, N-1$, $W_i(x_0) \geq 0$ and $W_i(x_N) \geq 0$, $i = 1, 2$. Then $W_i(x_j) \geq 0$, $\forall x_j \in \overline{\Omega}^N$, $i = 1, 2$.*

Proof Using the basic idea used in [12] for the continuous maximum principle and the discrete test functions given by, $S_1(x_j) = S_2(x_j) = 1 + x_j, \forall x_j \in \overline{\Omega}^N$, the above theorem can be proved. □

The following discrete stability result can be proved by using the above discrete maximum principle.

Corollary 2 (Discrete Stability Result) *Let $\overline{W} = (W_1,\ W_2)$ be any mesh function. Then*

$$|\ W_i(x_j)\ | \leq C\ \max\left\{\max_{k=1,2}\{|\ W_k(x_0)\ |\},\ \max_{k=1,2}\{|\ W_k(x_N)\ |\},\ \frac{1}{\alpha}\max_{k=1,2}\{\sup_{\xi\in\Omega^N} |P_k^{*N}\overline{W}(\xi)|\}\right\},$$
$$\forall\, x_j \in \overline{\Omega}^N,\ i = 1, 2,$$

where $\Omega^N = \overline{\Omega}^N \setminus \{x_0, x_N\}$.

6.2 Error Estimate

Using the Theorem 5.1 of [20] and the procedure adapted in the above Sect. 5.2, we derive the following result

Theorem 6 *Let $\overline{u}$ and $\overline{U}^{[n]}$ be the solutions of the problems (2) and (17) respectively, $\sqrt{\varepsilon} \le CN^{-1}$. Then we have*
(i) Shishkin mesh:

$$\|u_i - U_i^{[n]}\|_{\overline{\Omega}^N} \le CN^{-2}\ln^2 N, \quad i = 1, 2, \quad \textit{provided that} \quad n \ge \frac{\ln(N^{-2}\ln^2 N)}{\ln \gamma}, \quad \gamma = \frac{\beta}{\alpha}.$$

and
(ii) BS mesh:

$$\|u_i - U_i^{[n]}\|_{\overline{\Omega}^N} \le CN^{-2}, \quad i = 1, 2, \quad \textit{provided that} \quad n \ge \frac{\ln(N^{-2})}{\ln \gamma}, \quad \gamma = \frac{\beta}{\alpha}.$$

Proof Consider the problem (9). From (17) we have

$$\begin{cases} P_i^{*N}\overline{U}_j^{[1]} = \begin{cases} \overline{h}_j(f_{i,j} - \sum_{k=1}^{2} b_{ik,j}\phi_k(x_j - 1)), & j = 1, \cdots, \frac{N}{2} - 1, \\ \frac{\overline{h}_j}{2}(f_{i,j-1} - \sum_{k=1}^{2} b_{ik,j-1}\phi_k(x_{j-1} - 1) \\ +f_{i,j+1} - \sum_{k=1}^{2} b_{ik,j+1}u_k^{(0)}(x_{j+1} - 1)), & j = \frac{N}{2}, \\ \overline{h}_j(f_{i,j} - \sum_{k=1}^{2} b_{ik,j}u_k^{(0)}(x_j - 1)), & j = \frac{N}{2} + 1, \cdots, N - 1, \end{cases} \\ U_{i,0}^{[1]} = U_{i,N}^{[1]} = 0, \quad i = 1, 2, \end{cases} \tag{18}$$

Applying the error estimate result from [20] to the problems (9)–(18) we have

$$|u_i^{(1)}(x_j) - U_{i,j}^{[1]}| \le \begin{cases} CN^{-2}\ln^2 N, & \text{for Shishkin mesh,} \\ CN^{-2}, & \text{for BS mesh.} \end{cases} \tag{19}$$

As done in Sect. 5, using (6) and (19) we get

$$\|u_i - U_i^{[1]}\|_{\overline{\Omega}^N} \le \begin{cases} C\left(\frac{\beta}{\alpha}\right) + CN^{-2}\ln^2 N & \text{for Shishkin mesh,} \\ C\left(\frac{\beta}{\alpha}\right) + CN^{-2} & \text{for BS mesh.} \end{cases}$$

Now consider the problem (12). The scheme (17) corresponding to this problem is

$$\begin{cases} P_i^{*N}\overline{U}_j^{*[2]} = \begin{cases} \overline{h}_j(f_{i,j} - \sum_{k=1}^{2} b_{ik,j}\phi_k(x_j - 1)), & j = 1, \cdots, \frac{N}{2} - 1, \\ \frac{\overline{h}_j}{2}(f_{i,j-1} - \sum_{k=1}^{2} b_{ik,j-1}\phi_k(x_{j-1} - 1) \\ +f_{i,j+1} - \sum_{k=1}^{2} b_{ik,j+1}u_k^{(1)}(x_{j+1} - 1)), & j = \frac{N}{2}, \\ \overline{h}_j(f_{i,j} - \sum_{k=1}^{2} b_{ik,j}u_k^{(1)}(x_j - 1)), & j = \frac{N}{2} + 1, \cdots, N - 1, \end{cases} \\ U_{i,0}^{*[2]} = U_{i,N}^{*[2]} = 0, \quad i = 1, 2, \end{cases} \tag{20}$$

Applying the error estimate result from [20] to the problems (12) and (20) we have

$$|u_i^{(2)}(x_j) - U_{i,j}^{[2]}| \leq \begin{cases} CN^{-2}\ln^2 N, & \text{for Shishkin mesh,} \\ CN^{-2}, & \text{for BS mesh,} \quad j = 0, 1, \cdots, N, \quad i = 1, 2. \end{cases} \tag{21}$$

Replacing $u_k^{(1)}(x_j - 1)$ by its numerical solution we get

$$\begin{cases} P_i^{*N}\overline{U}_j^{[2]} = \begin{cases} \overline{h}_j(f_{i,j} - \sum_{k=1}^{2} b_{ik,j}\phi_k(x_j - 1)), & j = 1, \cdots, \frac{N}{2} - 1, \\ \frac{\overline{h}_j}{2}(f_{i,j-1} - \sum_{k=1}^{2} b_{ik,j-1}\phi_k(x_{j-1} - 1) \\ +f_{i,j+1} - \sum_{k=1}^{2} b_{ik,j+1}U_{k,1}^{[1]}), & j = \frac{N}{2}, \\ \overline{h}_j(f_{i,j} - \sum_{k=1}^{2} b_{ik,j}U_{k,j-\frac{N}{2}}^{[1]}), & j = \frac{N}{2} + 1, \cdots, N - 1, \end{cases} \\ U_{i,0}^{[2]} = U_{i,N}^{[2]} = 0, \quad i = 1, 2, \end{cases} \tag{22}$$

Applying Corollary 2 for $\overline{Z}^{[2]} = (Z_1^{[2]}, Z_2^{[2]}), Z_{i,j}^{[2]} = U^{*}{}_{i,j}^{[2]} - U_{i,j}^{[2]}$ where $\overline{Z}^{[2]}$ satisfies the following discrete problem

$$\begin{cases} P_i^{*N}\,\overline{Z}_j^{[2]} = \begin{cases} 0, & j = 1, \cdots, \frac{N}{2} - 1, \\ -\frac{h_j}{2}\left(\sum_{k=1}^{2} b_{ik,j+1}\left(u_k^{(1)}(x_{j+1} - 1) - U_{k,1}^{[1]}\right)\right), & j = \frac{N}{2}, \\ -h_j\left(\sum_{k=1}^{2} b_{ik,j}\left(u_k^{(1)}(x_j - 1) - U_{k,j-\frac{N}{2}}^{[1]}\right)\right), & j = \frac{N}{2} + 1, \cdots, N - 1, \end{cases} \\ Z_{i,0}^{[2]} = 0, \quad Z_{i,N}^{[2]} = 0, \quad i = 1, 2, \end{cases}$$

we get

$$|U^{*}{}_{i,j}^{[2]} - U_{i,j}^{[2]}| \leq \begin{cases} \frac{\beta}{\alpha}CN^{-2}\ln^2 N < CN^{-2}\ln^2 N, & \text{for Shishkin mesh,} \\ \frac{\beta}{\alpha}CN^{-2} < CN^{-2}, & \text{for BS mesh, } j = 0, 1, \cdots, N, \quad i = 1, 2. \end{cases}$$

The proof of Theorem 4 can be used to show that

$$\|u_i - U_i^{[2]}\|_{\overline{\Omega}^N} \leq \begin{cases} C\left(\frac{\beta}{\alpha}\right)^2 + CN^{-2}\ln^2 N, & \text{for Shishkin mesh,} \\ C\left(\frac{\beta}{\alpha}\right)^2 + CN^{-2}, & \text{for BS mesh,} \quad i = 1, 2. \end{cases}$$

Continuing this process one can prove that

$$\|u_i - U_i^{[n]}\|_{\overline{\Omega}^N} \leq \begin{cases} C\left(\frac{\beta}{\alpha}\right)^n + CN^{-2}\ln^2 N, & \text{for Shishkin mesh,} \\ C\left(\frac{\beta}{\alpha}\right)^n + CN^{-2}, & \text{for BS mesh,} \quad i = 1, 2, \quad n > 2. \end{cases}$$

Since $\frac{\ln(N^{-2}\ln^2 N)}{\ln \gamma} \leq n$ and $\frac{\ln(N^{-2})}{\ln \gamma} \leq n$ for Shishkin mesh and BS mesh respectively and $\gamma = \frac{\beta}{\alpha}$, we have $(\frac{\beta}{\alpha})^n \leq N^{-2}\ln^2 N$ and $(\frac{\beta}{\alpha})^n \leq N^{-2}$. Finally we have

$$\|u_i - U_i^{[n]}\|_{\overline{\Omega}^N} \leq \begin{cases} CN^{-2}\ln^2 N, & \text{for Shishkin mesh,} \\ CN^{-2}, & \text{for BS mesh,} \quad i = 1, 2. \end{cases}$$

Hence the proof of the theorem. □

7 Numerical Experiments

In all the examples considered below, the number of iterations used is the smallest integer n such that

$$n \geq \frac{\ln(N^{-1}\ln N)}{\ln \gamma}, \quad \gamma = \frac{\beta}{\alpha}, \tag{23}$$

for the scheme given in Sect. 5.1. A similar criteria is used for the scheme given in Sect. 6.1.

Let $U_{1,j}^N$ and $U_{2,j}^N$ denote $U_{1,j}^{[n]}$ and $U_{2,j}^{[n]}$ which met stopping criterion (23) for N points.

To find the maximum error, rate of convergence and error constant to the computed solution, the following double mesh principle is used. For this we put

$$D_{i,\varepsilon}^N = \max_{0\leq j\leq N} \mid U_{i,j}^N - U_{i,2j}^{2N} \mid, \quad i = 1, 2,$$

where $U_{i_j}^N$ and $U_{i_{2j}}^{2N}$ are the jth and $2j$th components of the numerical solutions on meshes of N and $2N$ points respectively. The uniform error and rate of convergence as

$$D_i^N = \max_{\varepsilon} D_{i,\varepsilon}^N, \; p_i^N = \log_2\left(\frac{D_i^M}{D_i^{2N}}\right), \; i = 1, 2.$$

The ranges of values of ε used to present the numerical results for the following examples are $\{2^{-15}, 2^{-14}, \cdots, 2^{-6}\}$ and $\{2^{-19}, 2^{-18}, \cdots, 2^{-15}\}$ for Shishkin mesh and BS mesh respectively.

Example 1 [12]

$$\begin{cases} -\varepsilon u_1''(x) + 11u_1(x) - (x^2+1)u_1(x-1) - (x+1)u_2(x-1) = \exp(x), & x \in \Omega, \\ -\varepsilon u_2''(x) + 16u_2(x) - xu_1(x-1) - xu_2(x-1) = \exp(x), & x \in \Omega, \\ u_1(x) = 1, \; x \in [-1, 0], \; u_1(2) = 1, \\ u_2(x) = 1, \; x \in [-1, 0], \; u_2(2) = 1, \end{cases} \tag{24}$$

Example 2

$$\begin{cases} -\varepsilon u_1''(x) + 2u_1(x) - u_2(x) - (x^2+1)u_1(x-1) - (x+1)u_2(x-1) = 1, \ x \in \Omega, \\ -\varepsilon u_2''(x) - u_1(x) + 2u_2(x) - xu_1(x-1) - xu_2(x-1) = 1, \ x \in \Omega, \\ u_1(x) = 1, \ x \in [-1, 0], \ u_1(2) = 1, \\ u_2(x) = 1, \ x \in [-1, 0], \ u_2(2) = 1, \end{cases} \tag{25}$$

Table 1 Numerical result for Example 1 using finite difference method (Shishkin mesh)

ε	Number of mesh points N					
	64	128	256	512	1024	2048
2^{-6}	5.1574e-2	3.2754e-2	2.1024e-2	1.2629e-2	7.3995e-3	4.2054e-3
Iterations	2	3	3	3	4	4
2^{-7}	5.1933e-2	3.2911e-2	2.1114e-2	1.2675e-2	7.4222e-3	4.2184e-3
Iterations	2	3	3	3	4	4
2^{-8}	5.2117e-2	3.2993e-2	2.1161e-2	1.2699e-2	7.4340e-3	4.2252e-3
Iterations	2	3	3	3	4	4
2^{-9}	5.2211e-2	3.3035e-2	2.1185e-2	1.2712e-2	7.4402e-3	4.2287e-3
Iterations	2	3	3	3	4	4
2^{-10}	5.2258e-2	3.3056e-2	2.1198e-2	1.2718e-2	7.4433e-3	4.2304e-3
Iterations	2	3	3	3	4	4
2^{-11}	5.2282e-2	3.3066e-2	2.1204e-2	1.2721e-2	7.4449e-3	4.2313e-3
Iterations	2	3	3	3	4	4
2^{-12}	5.2294e-2	3.3072e-2	2.1207e-2	1.2723e-2	7.4457e-3	4.2318e-3
Iterations	2	3	3	3	4	4
2^{-13}	5.2300e-2	3.3074e-2	2.1208e-2	1.2724e-2	7.4461e-3	4.2320e-3
Iterations	2	3	3	3	4	4
2^{-14}	5.2302e-2	3.3076e-2	2.1209e-2	1.2724e-2	7.4463e-3	4.2321e-3
Iterations	2	3	3	3	4	4
2^{-15}	5.2304e-2	3.3076e-2	2.1210e-2	1.2724e-2	7.4464e-3	4.2322e-3
Iterations	2	3	3	3	4	4
D_1^N	5.2304e-2	3.3076e-2	2.1210e-2	1.2724e-2	7.4464e-3	4.2322e-3
P_1^N	0.6611	0.6411	0.7371	0.7730	0.8151	–

Table 2 Numerical results for Example 1

N	64	128	256	512	1024	2048
Finite Difference scheme (Shishkin mesh)						
D_2^N	9.3629e-3	5.9954e-3	3.7002e-3	2.2098e-3	1.2801e-3	7.2479e-4
p_2^N	0.6431	0.6963	0.7437	0.7877	0.8206	–
Finite Element method (Shishkin mesh)						
D_1^N	7.9973e-3	4.3482e-3	1.9510e-3	8.3107e-4	3.1085e-4	1.0563e-4
p_1^N	0.8791	1.1562	1.2312	1.4188	1.5572	–
D_2^N	8.9543e-3	5.3117e-3	2.7851e-3	1.2847e-3	4.901e-4	1.7257e-4
p_2^N	0.7534	0.9314	1.1163	1.3903	1.5059	–
Finite Element method (BS mesh)						
D_1^N	2.3584e-2	6.5168e-3	1.7037e-3	4.3490e-4	1.0982e-4	2.7590e-5
p_1^N	1.8555	1.9355	1.9699	1.9855	1.9929	–
D_2^N	8.7156e-3	2.2031e-3	5.5483e-4	1.3916e-4	3.4829e-5	8.7112e-6
p_2^N	1.9840	1.9894	1.9952	1.9984	1.9993	–

Table 3 Numerical results for Example 2

N	64	128	256	512	1024	2048
Finite Difference scheme (Shishkin mesh)						
D_1^N	1.4468e-2	1.0713e-2	6.8360e-3	4.1030e-3	2.3105e-3	1.2826e-3
p_1^N	0.4335	0.6481	0.7365	0.8285	0.8491	–
D_2^N	1.2130e-2	7.9075e-3	4.9016e-3	2.9443e-3	1.6815e-3	9.4230e-4
p_2^N	0.6173	0.6900	0.7353	0.8082	0.8355	–
Finite Element method (Shishkin mesh)						
D_1^N	5.2581e-4	2.9272e-4	1.4352e-4	6.1086e-5	2.4814e-5	8.6222e-6
p_1^N	0.8450	1.0282	1.2323	1.2996	1.5250	–
D_2^N	5.0298e-4	2.7196e-4	1.3244e-4	6.0314e-5	2.3404e-5	8.0203e-6
p_2^N	0.8871	1.0381	1.1348	1.3657	1.5450	–
Finite Element method (BS mesh)						
D_1^N	1.8931e-3	5.0521e-4	1.2988e-4	3.2891e-5	8.2735e-6	2.0746e-6
p_1^N	1.9057	1.9596	1.9814	1.9911	1.9956	–
D_2^N	3.4511e-4	8.918e-5	2.2922e-5	5.8009e-6	1.4559e-6	3.4563e-7
p_2^N	1.9523	1.9600	1.9824	1.9944	2.0746	–

Tables 1, 2 and 3 present the values of D_i^N, P_i^N, $i = 1, 2$ for the above Examples 1–2 respectively. Fig. 1 represents the numerical solution of the problem stated in Examples 1.

Fig. 1 Numerical solution of the above Example 1 for $\varepsilon = 2^{-14}$ and $N = 128$

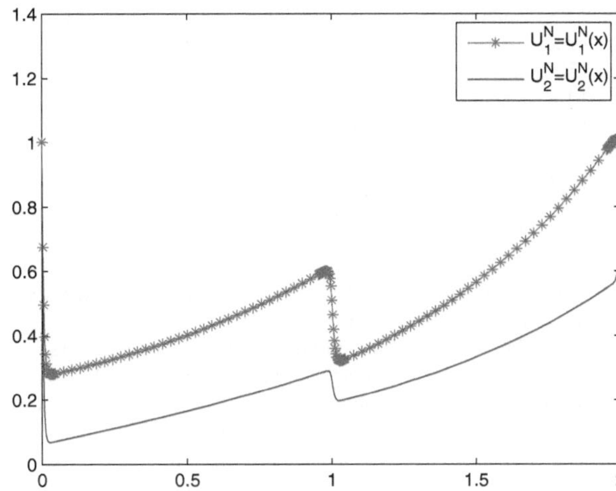

8 Discussion

A numerical method is suggested in this paper for a system of singularly perturbed differential equations of reaction-diffusion type with negative shift. This method is based on an iterative procedure, new and easy to implement. A numerical method namely initial value technique for the above problem (1) is proposed in [12]. The error estimate derived by them depends on the parameter ε though they arrived at the maximum order of convergence as almost two. In our method though the error estimate depends on the number of iterations, this number grows slowly with N. Further the bound is independent of the parameter ε. Further using a finite element method on BS mesh, we are able to improve the order of the convergence by two.

Acknowledgments The first author thank the Department of Science and Technology, Government of India, for their financial support under the DST/INSPIRE Fellowship/2014/[IF140503].

References

1. A. Longtin, J.G. Milton, Complex oscillations in the human pupil light reflex with mixed and delayed feedback. Math. Biosci. **90**, 183–199 (1988)
2. M.W. Derstine, H.M. Gibbs, F.A. Hopf, D.L. Kaplan, Bifurcation gap in a hybrid optically bistable system. Phys. Rev. A **26**(6), 3720–3722 (1982)
3. V.Y. Glizer, Asymptotic analysis and solution of a finite-horizon H_∞ control problem for singularly-perturbed linear systems with small state delay. J. Optim. Theory Appl. **117**(2), 295–325 (2003)
4. F. Erdogan, An exponentially fitted method for singularly perturbed delay differential equations. Adv. Differ. Eq. **2009**, 781579 (2009). doi:10.1155/2009/781579

5. M.K. Kadalbajoo, K.K. Sharma, Numerical treatment of boundary value problems for second order singularly perturbed delay differential equations. Comput. Appl. Math. **24**(2), 151–172 (2005)
6. M.K. Kadalbajoo, K.K. Sharma, Parameter-Uniform fitted mesh method for singularly perturbed delay differential equations with layer behavior. Electron. Trans. Numer. Anal. **23**, 180–201 (2006)
7. M.K. Kadalbajoo, K.K. Sharma, A numerical method based on finite difference for boundary value problems for singularly perturbed delay differential equations. Appl. Math. Comput. **197**, 692–707 (2008)
8. J. Mohapatra, S. Natesan, Uniform convergence analysis of finite difference scheme for singularly perturbed delay differential equation on an adaptively generated grid. Numer. Math.: Theory, Methods Appl. **3**, 1–22 (2010)
9. V. Subburayan, N. Ramanujam, An initial value technique for singularly perturbed reaction-diffusion problems with a negative shift. Novi Sad J. Math. **43**(2), 67–80 (2013)
10. V. Subburayan, N. Ramanujam, *An Initial Value Method for Singularly Perturbed Third order Delay Differential Equations*, International Conference on Mathematical Sciences (ICMS), Sathyabhama university, Chennai, India, July 17–19, 2014, pp. 221–229
11. V. Subburayan, N. Ramanujam, An asymptotic numerical method for singularly perturbed convection-diffusion problems with a negative shift. Neural, Parallel Scient. Comput. **21**, 431–446 (2013)
12. V. Subburayan, N. Ramanujam, An initial value method for singularly perturbed system of reaction-diffusion type delay differential equations. J. KSIAM. **17**(4), 221–237 (2013)
13. V. Subburayan, N. Ramanujam, An initial value technique for singularly perturbed convection-diffusion problems with a negative shift. J. Optim. Theory Appl. **158**, 234–250 (2013)
14. V. Subburayan, N. Ramanujam, Asymptotic initial value technique for singularly perturbed convection diffusion delay problems with boundary and weak interior layers. Appl. Math. Lett. **25**, 2272–2278 (2012)
15. S. Nicaise, C. Xenophontos, Robust approximation of singularly perturbed delay differential equations by the hp finite element method. Comput. Meth. Appl. Math. **13**(1), 21–37 (2013)
16. H. Zarin, On discontinuous Galerkin finite element method for singularly perturbed delay differential equations. Appl. Math. Lett. **38**, 27–32 (2014)
17. P. Chocholaty, L. Slahor, A numerical method to boundary value problems for second order delay differential equations. Numerische Mathematik **33**, 69–75 (1979)
18. C.G. Lange, R.M. Miura, Singularly perturbation analysis of boundary-value problems for differential-difference equations. SIAM. J. Appl. Math. **42**(3), 502–531 (1982)
19. A. Tamilselvan, N. Ramanujam, V. Shanthi, A numerical method for singularly perturbed weakly coupled system of two second order ordinary differential equations with discontinuous source term. J. Comput. Appl. Math. **202**, 203–216 (2007)
20. A. Ramesh Babu, N. Ramanujam, An almost second order FEM for a weakly coupled system of two singularly perturbed differential equations of reaction-diffusion type with discontinuous source term. Neural, Parallel, Scient. Comput. **17**, 147–168 (2009)

Numerical Method for a Singularly Perturbed Boundary Value Problem for a Linear Parabolic Second Order Delay Differential Equation

Parthiban Swaminathan, Valarmathi Sigamani and Franklin Victor

Abstract A singularly perturbed boundary value problem for a linear parabolic second order delay differential equation of reaction-diffusion type is considered. As the highest order space derivative is multiplied by a singular perturbation parameter, the solution exhibits boundary layers. Also, the delay term that occurs in the space variable gives rise to interior layers. A numerical method which uses classical finite difference scheme on a Shishkin piecewise uniform mesh is suggested to approximate the solution. The method is proved to be first order convergent uniformly with respect to the singular perturbation parameter. Numerical illustrations are also presented.

Keywords Singular perturbation problems · Boundary layers · Parabolic delay-differential equations · Finite difference scheme · Shishkin mesh · Parameter uniform convergence

1 Introduction

Singularly Perturbed Delay Differential Equations (SPDDE) are being used to model many practical problems in different branches of science such as Biomathematics, Control Theory, Ecology, etc.

This work is confined to the class of boundary value problems for linear second order singularly perturbed delay differential equations of parabolic type. As singular perturbation parameter (ε) multiplies the highest order derivative term of the equation, boundary layers appear on the lateral sides of the rectangular domain. These boundary layers are of parabolic type since the characteristic of the reduced equation

P. Swaminathan · V. Sigamani (✉) · F. Victor
Department of Mathematics, Bishop Heber College, Tiruchirappalli, Tamil Nadu, India
e-mail: valarmathi07@gmail.com

P. Swaminathan
e-mail: parthitsaminathan@gmail.com

F. Victor
e-mail: franklinvicto@gmail.com

V. Sigamani et al. (eds.), *Differential Equations and Numerical Analysis*,
Springer Proceedings in Mathematics & Statistics 172,
DOI 10.1007/978-81-322-3598-9_7

(by putting $\varepsilon = 0$) are parallel to the boundary. Further, due to a lag which is assumed in the space variable, interior layers also arise. Since the classical finite difference schemes on uniform meshes fail to give good approximations, it is essential to construct a numerical method which is robust and layer resolving. General introduction to finite difference methods to solve singularly perturbed differential equations of parabolic reaction-diffusion type can be found in [1–5]. In [6–8] numerical methods for singularly perturbed delay differential equations are found. In [7], a numerical method comprising a standard finite difference operator (centered in space, implicit in time) on a rectangular piecewise uniform fitted mesh condensing at the boundary layers is established to solve singularly perturbed delay differential equations of parabolic type. The delay term is assumed to occur in time variable. This method is proved to be uniformly convergent with respect to the parameter (ε).

Here in this paper a numerical method which uses standard finite difference scheme on a Shishkin piecewise uniform mesh is constructed. It is proved that the numerical approximations obtained by this method converge to the exact solution uniformly for all the values of the parameter in the maximum norm. The plan of the paper is as follows. In Sect. 2, the problem is defined and existence and regularity of the solution of the problem are discussed. In Sect. 3, the maximum principle for the differential operator is proved and consequently the stability result is established. And also standard estimates of the derivatives of the solution are presented. Further, improved estimates for the derivatives of components of the solution are presented. In Sect. 4, piecewise-uniform Shishkin meshes are introduced and in Sect. 5, the discrete problem is defined and the discrete maximum principle and the discrete stability properties are established. In Sect. 6, numerical analysis is presented and the error bounds are established. In Sect. 7, numerical illustrations are presented.

2 The Continuous Problem

A singularly perturbed boundary value problem for a linear parabolic second order delay differential equation of reaction—diffusion type is considered as follows

$$Lu(x,t) = \frac{\partial u}{\partial t}(x,t) - \varepsilon\frac{\partial^2 u}{\partial x^2}(x,t) + a(x,t)u(x,t) + b(x,t)u(x-1,t) = f(x,t) \text{ on } \Omega,$$
$$u \text{ given on } \Gamma,\ u(x,t) = \chi(x,t),\ (x,t) \in [-1,0]\times[0,T], \tag{1}$$

where $\Omega = \{(x,t) : 0 < x < 2, 0 < t \le T\}$, $\tilde{\Omega} = ((0,1-)\times(0,T]) \cup ((1+,2)\times(0,T])$, $\overline{\tilde{\Omega}} = ([0,1-]\times[0,T]) \cup ([1+,2]\times[0,T])$ $\overline{\Omega}{=}\Omega\cup\Gamma$, $\Gamma = \Gamma_L \cup \Gamma_B \cup \Gamma_R$ with $u(0,t){=}\chi(0,t)$ on $\Gamma_L = \{(0,t) : 0 \le t \le T\}$, $u(x,0) = \phi_B(x)$ on $\Gamma_B = \{(x,0) : 0 \le x \le 2\}$, and $u(2,t) = \phi_R(t)$ on $\Gamma_R = \{(2,t) : 0 \le t \le T\}$. The functions χ, ϕ_B, and ϕ_R are assumed to be sufficiently smooth. It is to be noted that the domain of the operator L is $\mathscr{M}_\lambda(\Omega) = \{\psi : \frac{\partial\psi}{\partial t}, \frac{\partial^2\psi}{\partial x^2} \text{ exist on } \Omega\}$.

Standard theoretical results on the solutions of (1) are stated, without proof, in this section. See [2, 9, 10] for more details.

For all $(x, t) \in [0, 2] \times [0, T]$, it is assumed that $a(x, t)$ and $b(x, t)$ satisfy

$$a(x, t) + b(x, t) > 2\alpha, \quad \text{for some real number } \alpha > 0 \tag{2}$$

$$\text{and} \quad b(x, t) < 0 \tag{3}$$

The problem (1) can be rewritten as,

$$L_1 u(x, t) = \frac{\partial u}{\partial t}(x, t) - \varepsilon \frac{\partial^2 u}{\partial x^2}(x, t) + a(x, t)u(x, t) = g(x, t), \text{ on } \Omega_1 = (0, 1) \times (0, T] \tag{4}$$

where $g(x, t) = f(x, t) - b(x, t)\chi(x - 1, t)$

$$L_2 u(x, t) = \frac{\partial u}{\partial t}(x, t) - \varepsilon \frac{\partial^2 u}{\partial x^2}(x, t) + a(x, t)u(x, t) + b(x, t)u(x - 1, t) = f(x, t), \quad \text{on } \Omega_2 = (1, 2) \times (0, T] \tag{5}$$

$u(0, t) = \chi(0, t), u(x, 0) = \phi_B(x)$ on $\Gamma_{B_1} = \{(x, 0) : 0 \le x \le 1-\}$,$u(1-, t) = u(1+, t)$, $u'(1-, t) = u'(1+, t), u(x, 0) = \phi_B(x)$ on $\Gamma_{B_2} = \{(x, 0) : 1+ \le x \le 2\}, u(2, t) = \phi_R(t)$ on Γ_R.

The reduced problem corresponding to (4)–(5) is defined by

$$\frac{\partial u_0}{\partial t}(x, t) + a(x, t)u_0(x, t) = g(x, t), \text{ on } (0, 1) \times (0, T]$$
$$u_0(x, 0) = \phi_B(x), 0 \le x \le 1 \tag{6}$$

$$\frac{\partial u_0}{\partial t}(x, t) + a(x, t)u_0(x, t) + b(x, t)u_0(x - 1, t) = f(x, t), \text{ on } (1, 2) \times (0, T]$$
$$u_0(x, 0) = \phi_B(x), 1+ \le x \le 2. \tag{7}$$

In general as $u_0(x, t)$ need not satisfy $u_0(0, t) = u(0, t)$ and $u_0(2, t) = u(2, t)$, the solution $u(x, t)$ exhibits boundary layers at $x = 0$ and $x = 2$. In addition to that, as $u_0(1-, t)$ need not be equal to $u_0(1+, t)$, the solution $u(x, t)$ exhibits interior layers at $x = 1$.

For any function y on a domain D the following norm is introduced: $\| y \|_D = \sup_{(x,t)\in D} |y(x, t)|$. If $D = \overline{\Omega}$, the subscript is dropped.

In a compact domain D a function is said to be Hölder continuous of degree $\lambda, 0 < \lambda \le 1$, if, for all $(x_1, t_1), (x_2, t_2) \in D$,

$$|u(x_1, t_1) - u(x_2, t_2)| \le C(|x_1 - x_2|^2 + |t_1 - t_2|)^{\lambda/2}.$$

The set of Hölder continuous functions forms a normed linear space $C^0_\lambda(D)$ with the norm

$$||u||_{\lambda,D} = ||u||_D + \sup_{(x_1,t_1),(x_2,t_2)\in D} \frac{|u(x_1,t_1) - u(x_2,t_2)|}{(|x_1 - x_2|^2 + |t_1 - t_2|)^{\lambda/2}}.$$

For each integer $k \geq 1$, the subspaces $C^k_\lambda(D)$ of $C^0_\lambda(D)$, which contain functions having Hölder continuous derivatives, are defined as follows

$$C^k_\lambda(D) = \{u : \frac{\partial^{l+m} u}{\partial x^l \partial t^m} \in C^0_\lambda(D) \text{ for } l, m \geq 0 \text{ and } 0 \leq l + 2m \leq k\}.$$

The norm on $C^0_\lambda(D)$ is taken to be $||u||_{\lambda,k,D} = \max_{0\leq l+2m\leq k} ||\frac{\partial^{l+m} u}{\partial x^l \partial t^m}||_{\lambda,D}$.

Sufficient conditions for the existence, uniqueness and regularity of solution of (1) are given in the following theorem.

Theorem 1 *Assume that* $a, b, f \in C^2_\lambda(\overline{\Omega})$, $\chi \in C^1_\lambda([-1,0] \times [0,T])$, $\phi_B \in C^2(\Gamma_B)$, $\phi_R \in C^1(\Gamma_R)$ *and that the following compatibility conditions are fulfilled at the corners* $(0,0)$ *and* $(2,0)$ *of* Γ,

$$\phi_B(0) = \chi(0,t) \text{ and } \phi_B(2) = \phi_R(0). \tag{8}$$

$$\begin{aligned}
&\frac{\partial \chi}{\partial t}(0,0) - \varepsilon \frac{d^2\phi_B}{dx^2}(0) + a(0,0)\phi_B(0) + b(0,0)\chi(0,0) = f(0,0),\\
&\frac{d\phi_R}{dt}(0) - \varepsilon \frac{d^2\phi_B}{dx^2}(2) + a(2,0)\phi_B(2) + b(2,0)\phi_B(2) \quad = f(2,0)
\end{aligned} \tag{9}$$

and

$$\begin{aligned}
\frac{\partial^2 \chi}{\partial t^2}(0,0) = \varepsilon^2 \frac{d^4\phi_B}{dx^4}(0) &- \varepsilon\{\frac{\partial^2 a}{\partial x^2}(0,0)\phi_B(0) - 2\frac{\partial a}{\partial x}(0,0)\phi_B(0) + a(0,0)\frac{d^2\phi_B}{dx^2}(0)\\
&+ \frac{\partial^2 b}{\partial x^2}(0,0)\chi(-1,0) - 2\frac{\partial b}{\partial x}(0,0)\chi(-1,0) + b(0,0)\frac{\partial^2 \chi}{\partial x^2}(-1,0)\}\\
&- \{\frac{\partial a}{\partial t}(0,0)\phi_B(0) + b(0,0)\frac{\partial \chi}{\partial t}(-1,0) + \frac{\partial b}{\partial t}(0,0)\chi(-1,0)\}\\
&+ \varepsilon \frac{\partial^2 f}{\partial x^2}(0,0) + \frac{\partial f}{\partial t}(0,0),
\end{aligned} \tag{10}$$

$$\frac{d^2\phi_R}{dt^2}(0) = \varepsilon^2 \frac{d^4\phi_B}{dx^4}(2) - \varepsilon\{\frac{\partial^2 a}{\partial x^2}(2,0)\phi_B(2) - 2\frac{\partial a}{\partial x}(2,0)\phi_B(2) + a(2,0)\frac{d^2\phi_B}{dx^2}(2)$$

$$+\frac{\partial^2 b}{\partial x^2}(2,0)\phi_B(1) - 2\frac{\partial b}{\partial x}(2,0)\phi_B(1) + b(2,0)\frac{d^2\phi_B}{dx^2}(1)\}$$
$$-\{\frac{\partial a}{\partial t}(2,0)\phi_B(2) + b(2,0)\frac{d\phi_B}{dt}(1) + \frac{\partial b}{\partial t}(2,0)\phi_B(1)\}$$
$$+\varepsilon\frac{\partial^2 f}{\partial x^2}(2,0) + \frac{\partial f}{\partial t}(2,0). \tag{11}$$

Then there exists a unique solution u of (1) satisfying $u \in \mathscr{C} = C^0_\lambda([0,2]\times[0,T]) \cap C^1_\lambda((0,2)\times(0,T]) \cap C^4_\lambda(\bar{\tilde{\Omega}})$.

It is assumed throughout the paper that all of the assumptions (2), (3), (8), (9), (10) and (11) of this section hold. Furthermore, C denotes a generic positive constant, which is independent of x, t and of all singular perturbation and discretization parameters.

3 Analytical Results

The operator L satisfies the following maximum principle

Lemma 1 *Let ψ be any function in the domain of L such that $\psi \geq 0$ on Γ. Then $L\psi(x,t) \geq 0$ on Ω implies that $\psi(x,t) \geq 0$ on $\overline{\Omega}$.*

Proof Let x^*, t^* be such that $\psi(x^*,t^*) = \min_{\bar{\Omega}} \psi(x,t)$ and assume that the lemma is false. Then $\psi(x^*,t^*) < 0$. From the hypotheses we have $(x^*,t^*) \notin \Gamma$. With $\frac{\partial\psi}{\partial t}(x^*,t^*-) \leq 0$ and $\frac{\partial^2\psi}{\partial x^2}(x^*,t^*) \geq 0$,

$$\begin{aligned} L\psi(x^*,t^*) &= \frac{\partial\psi}{\partial t}(x^*,t^*) - \varepsilon\frac{\partial^2\psi}{\partial x^2}(x^*,t^*) + a(x^*,t^*)\psi(x^*,t^*) + b(x^*,t^*)\psi(x^*-1,t^*) \\ &< 0 \quad \text{as} \quad \psi(x^*-1,t^*) \geq \psi(x^*,t^*). \end{aligned}$$

This contradicts the assumption and proves the result of L.

As a consequence, the stability result for the problem (1) follows and is as stated in

Lemma 2 *If ψ is any function in the domain of L, then for all $(x,t) \in \overline{\Omega}$,*

$$|\psi(x,t)| \leq \max\left\{\|\ \psi\ \|_\Gamma, \frac{1}{\alpha}\ \|\ L\psi\ \|\right\}.$$

Proof Define the two functions

$$\theta^{\pm}(x,t) = \max\left\{\|\ \psi\ \|_\Gamma, \frac{1}{\alpha}\ \|\ L\psi\ \|\right\} \pm \psi(x,t).$$

Using the properties of $a(x,t)$ and $b(x,t)$, it is not hard to verify that $\theta^{\pm}(x,t) \geq 0$ for $(x,t) \in \Gamma$ and $L\theta^{\pm} \geq 0$ on Ω. It follows from Lemma 1 that $\theta^{\pm} \geq 0$ on $\bar{\Omega}$.

Standard estimates of the solution of (1) and its derivatives are contained in the following lemma.

Lemma 3 *Let u be the solution of (1). Then, for all* $(x,t) \in \overline{\tilde{\Omega}}$,

$$|\frac{\partial^k u}{\partial t^k}(x,t)| \le C(||u||_\Gamma + \textstyle\sum_{q=0}^{k} ||\frac{\partial^q f}{\partial t^q}||),\ k = 0, 1, 2$$
$$|\frac{\partial^k u}{\partial x^k}(x,t)| \le C\varepsilon^{\frac{-k}{2}}(||u||_\Gamma + ||f|| + ||\frac{\partial f}{\partial t}||),\ k = 1, 2$$
$$|\frac{\partial^k u}{\partial x^k}(x,t)| \le C\varepsilon^{\frac{-k}{2}}(||u||_\Gamma + ||f|| + ||\frac{\partial f}{\partial t}|| + ||\frac{\partial^2 f}{\partial t^2}|| + \varepsilon^{\frac{k-2}{2}}||\frac{\partial^{k-2} f}{\partial x^{k-2}}||),\ k = 3, 4$$
$$|\frac{\partial^k u}{\partial x^{k-1}\partial t}(x,t)| \le C\varepsilon^{\frac{1-k}{2}}(||u||_\Gamma + ||f|| + ||\frac{\partial f}{\partial t}|| + ||\frac{\partial^2 f}{\partial t^2}||),\ k = 2, 3.$$

Proof Following steps as in [2, 6], the required bounds are derived.

The Shishkin decomposition of the exact solution u of (1) is $u = v + w$ where the smooth component v is the solution of

$$L_1 v = g \text{ on } (0, 1-) \times (0, T],\ v(0,t) = u_0(0,t),\ v(x,0) = \phi_B(x),\ v(1-,t) = u_0(1-,t) \tag{12}$$

$$L_2 v = f \text{ on } (1+, 2) \times (0, T],\ v(2,t) = u_0(2,t),\ v(x,0) = \phi_B(x),\ v(1+,t) = u_0(1+,t) \tag{13}$$

and the singular component w is the solution of

$$\begin{aligned} &L_1 w = 0 \text{ on } (0,1) \times (0,T],\quad L_2 w = 0 \text{ on } (1,2) \times (0,T] \\ &\text{with } w(0,t) = u(0,t) - v(0,t),\ [w](1,t) = -[v](1,t),\ [w'](1,t) = -[v'](1,t), \\ &w(2,t) = u(2,t) - v(2,t),\ w(x,0) = 0. \end{aligned} \tag{14}$$

For convenience the left and right boundary layers of w are separated using

$$w(x,t) = w^L(x,t) + w^R(x,t) \tag{15}$$

with

$$\begin{aligned} w^L(x,t) &= w(0,t)w_1^L(x,t) + Aw_2^L(x,t), \\ \text{satisfying } L_1 w_1^L(x,t) &= 0,\ (x,t) \in (0,1) \times (0,T] \text{ with } w_1^L(0,t) = 1,\ w_1^L(1,t) = 0, \\ w_1^L(x,t) &= 0 \text{ on } (1,2] \times (0,T], \\ L_2 w_2^L(x,t) &= 0,\ (x,t) \in (1,2) \times (0,T] \text{ with } w_2^L(1,t) = 1,\ w_2^L(2,t) = 0, \\ w_2^L(x,t) &= 0 \text{ on } [0,1) \times (0,T], \\ \text{and } w^R(x,t) &= Bw_1^R(x,t) + w(2,t)w_2^R(x,t), \\ \text{satisfying } L_1 w_1^R(x,t) &= 0,\ (x,t) \in (0,1) \times (0,T] \text{ with } w_1^R(0,t) = 0,\ w_1^R(1,t) = 1, \\ w_1^R(x,t) &= 0 \text{ on } (1,2] \times (0,T], \\ L_2 w_2^R(x,t) &= 0,\ (x,t) \in (1,2) \times (0,T] \text{ with } w_2^R(1,t) = 0,\ w_2^R(2,t) = 1, \\ w_2^R(x,t) &= 0 \text{ on } [0,1) \times (0,T]. \end{aligned}$$

Here, A and B are constants to be chosen in such a way that the jump conditions at $x = 1$ are satisfied. Bounds on the smooth component and its derivatives are contained in the following lemma.

Lemma 4 *The smooth component v and its derivatives satisfy, for each $(x, t) \in \overline{\tilde{\Omega}}$,*
$|\frac{\partial^k v}{\partial t^k}(x, t)| \leq C, \text{ for } k = 0, 1, 2$
$|\frac{\partial^k v}{\partial x^k}(x, t)| \leq C(1 + \varepsilon^{1-\frac{k}{2}}), \text{ for } k = 0, 1, 2, 3, 4$
$|\frac{\partial^k v}{\partial x^{k-1}\partial t}(x, t)| \leq C, \text{ for } k = 2, 3.$

Proof Using the procedure adopted in [2, 6], the above bounds are derived.

4 Improved Estimates

The layer functions $B_1^L, B_1^R, B_2^L, B_2^R, B_1, B_2$, associated with the solution u, are defined by

$$B_1^L(x) = e^{-x\frac{\sqrt{\alpha}}{\sqrt{\varepsilon}}},\ B_1^R(x) = e^{-(1-x)\frac{\sqrt{\alpha}}{\sqrt{\varepsilon}}},\ B_1(x) = B_1^L(x) + B_1^R(x), \text{ on } [0, 1] \times [0, T],$$

$$B_2^L(x) = e^{-(x-1)\frac{\sqrt{\alpha}}{\sqrt{\varepsilon}}},\ B_2^R(x) = e^{-(2-x)\frac{\sqrt{\alpha}}{\sqrt{\varepsilon}}},\ B_2(x) = B_2^L(x) + B_2^R(x), \text{ on } [1, 2] \times [0, T].$$

Bounds on the singular components w^L and w^R of $u(x, t)$ and their derivatives are contained in the following lemma.

Lemma 5 *There exists a constant C, such that, for $(x, t) \in [0, 1-] \times [0, T]$,*
$|\frac{\partial^k w^L}{\partial t^k}(x, t)| \leq C B_1^L(x), \text{ for } k = 0, 1, 2,$
$|\frac{\partial^k w^L}{\partial x^k}(x, t)| \leq C\frac{B_1^L(x)}{\varepsilon^{\frac{k}{2}}}, \text{ for } k = 0, 1, 2, 3,$
$|\frac{\partial^k w^L}{\partial x^{k-1}\partial t}(x, t)| \leq C\frac{B_1^L(x)}{\varepsilon^{\frac{k}{2}}}, \text{ for } k = 0, 1, 2, 3.$

and for $(x, t) \in [1+, 2] \times [0, T]$,

$|\frac{\partial^k w^L}{\partial t^k}(x, t)| \leq C B_2^L(x), \text{ for } k = 0, 1, 2,$
$|\frac{\partial^k w^L}{\partial x^k}(x, t)| \leq C\frac{B_2^L(x)}{\varepsilon^{\frac{k}{2}}}, \text{ for } k = 0, 1, 2, 3,$
$|\frac{\partial^k w^L}{\partial x^{k-1}\partial t}(x, t)| \leq C\frac{B_2^L(x)}{\varepsilon^{\frac{k}{2}}}, \text{ for } k = 0, 1, 2, 3.$
Analogous results hold for w^R and its derivatives.

Proof The required bounds are derived following steps as in [2, 6].

Using the procedure adopted in [2, 6], sharper estimates of the smooth component derived and presented in the following lemma.

Lemma 6 *The smooth component* v *of the solution* u *of (1) satisfies, for* $(x,t) \in [0,1-] \times [0,T]$,

$|\frac{\partial v^k}{\partial x^k}(x,t)| \le C\,(1 + B_1(x))$, for $k = 0, 1, 2$ and $|\frac{\partial v^3}{\partial x^3}(x,t)| \le C\left(1 + \dfrac{B_1(x)}{\sqrt{\varepsilon}}\right)$.

and for $(x,t) \in [1+,2] \times [0,T]$,

$|\frac{\partial v^k}{\partial x^k}(x,t)| \le C\,(1 + B_2(x))$, for $k = 0, 1, 2$ and $|\frac{\partial v^3}{\partial x^3}(x,t)| \le C\left(1 + \dfrac{B_2(x)}{\sqrt{\varepsilon}}\right)$.

5 The Shishkin Mesh

A piecewise uniform Shishkin mesh with $M \times N$ mesh-intervals is now constructed. Let $\Omega_t^M = \{t_k\}_{k=1}^M$, $\Omega_x^N = \{x_j\}_{j=1}^{N-1}$, $\Omega^{M,N} = \Omega_t^M \times \Omega_x^N$, $\overline{\Omega}_t^M = \{t_k\}_{k=0}^M$, $\overline{\Omega}_x^N = \{x_j\}_{j=0}^N$, $\overline{\Omega}^{M,N} = \overline{\Omega}_t^M \times \overline{\Omega}_x^N$, $\Omega_x^{-N} = \{x_j\}_{j=1}^{\frac{N}{2}-1}$, $\Omega_x^{+N} = \{x_j\}_{j=\frac{N}{2}+1}^{N-1}$, $\Omega^{-M,N} = \Omega_t^M \times \Omega_x^{-N}$, $\Omega^{+M,N} = \Omega_t^M \times \Omega_x^{+N}$, $\overline{\Omega}_x^{-N} = \{x_j\}_{j=0}^{\frac{N}{2}}$, $\overline{\Omega}_x^{+N} = \{x_j\}_{j=\frac{N}{2}}^N$, $\overline{\Omega}^{-M,N} = \overline{\Omega}_t^M \times \overline{\Omega}_x^{-N}$, $\overline{\Omega}^{+M,N} = \overline{\Omega}_t^M \times \overline{\Omega}_x^{+N}$ and $\Gamma^{M,N} = \Gamma \cap \overline{\Omega}^{M,N}$. The mesh $\overline{\Omega}_t^M$ is chosen to be a uniform mesh with M mesh-intervals on $[0,T]$. The mesh $\overline{\Omega}_x^N$ is chosen to be a piecewise-uniform mesh with N mesh-intervals on $[0,2]$. The interval $[0,1]$ is divided into 3 sub-intervals as follows

$$[0,\tau] \cup (\tau, 1-\tau] \cup (1-\tau, 1].$$

The parameter τ, which determine the points separating the uniform meshes, is defined by

$$\tau = \min\left\{\frac{1}{4}, \frac{\sqrt{\varepsilon}}{\sqrt{\alpha}} \ln N\right\}. \tag{16}$$

Then, on the sub-interval $(\tau, 1-\tau]$ a uniform mesh with $\frac{N}{4}$ mesh points is placed and on each of the sub-intervals $[0,\tau]$ and $(1-\tau, 1]$, a uniform mesh of $\frac{N}{8}$ mesh points is placed.

Similarly, the interval $(1,2]$ is also divided into 3 sub-intervals $(1, 1+\tau]$, $(1+\tau, 2-\tau]$and$(2-\tau, 2]$, using the same parameter τ. In particular, when the parameter τ takes on its lefthand value, the Shishkin mesh $\overline{\Omega}^N$ becomes a classical uniform mesh throughout from 0 to 2.

In practice, it is convenient to take

$$N = 8k,\ k \ge 2. \tag{17}$$

From the above construction of $\overline{\Omega}^N$, it is clear that the transition points $\{\tau, 1-\tau, 1+\tau, 2-\tau\}$ are the only points at which the mesh-size can change and that it does not necessarily change at each of these points. The following notations are introduced: $h_j = x_j - x_{j-1}, h_{j+1} = x_{j+1} - x_j$ and if $x_j = \tau$, then $h_j^- = x_j - x_{j-1}, h_j^+ = x_{j+1} - x_j$, $J = \{x_j : h_j^+ \neq h_j^-\}$.

6 The Discrete Problem

In this section, a classical finite difference operator with an appropriate Shishkin mesh is used to construct a numerical method for (1) which is shown later to be essentially first order parameter-uniform convergent.
The discrete two-point boundary value problem is now defined on any mesh by the finite difference method

$$L^{M,N}U(x_j,t_k) = D_t^- U(x_j,t_k) - \varepsilon\delta_x^2 U(x_j,t_k) + a(x_j,t_k)U(x_j,t_k) \\ +b(x_j,t_k)U(x_j-1,t_k) = f(x_j,t_k) \text{ on } \Omega^{M,N} \quad (18)$$

$$U = u \text{ on } \Gamma^{M,N}$$

The problem (18) can be rewritten as

$$L_1^{M,N}U(x_j,t_k) = D_t^- U(x_j,t_k) - \varepsilon\delta_x^2 U(x_j,t_k) + a(x_j,t_k)U(x_j,t_k) = g(x_j,t_k) \text{ on } \Omega^{-M,N}$$
$$\text{where } g(x_j,t_k) = f(x_j,t_k) - b(x_j,t_k)\chi(x_j-1,t_k)$$
$$L_2^{M,N}U(x_j,t_k) = D_t^- U(x_j,t_k) - \varepsilon\delta_x^2 U(x_j,t_k) + a(x_j,t_k)U(x_j,t_k) + b(x_j,t_k)U(x_j-1,t_k) \\ = f(x_j,t_k) \text{ on } \Omega^{+M,N} \quad (19)$$

$$U = u \text{ on } \Gamma^{M,N}$$

$$D_x^- U(x_{\frac{N}{2}},t_k) = D_x^+ U(x_{\frac{N}{2}},t_k)$$

This is used to compute numerical approximations to the exact solution of (1). The following discrete results are analogous to those for the continuous case.

Lemma 7 *For any mesh function $\Psi(x_j,t_k), 0 \le j \le N, 0 \le k \le M$, the inequalities $\Psi \ge 0$ on $\Gamma^{M,N}, L_1^{M,N}\Psi(x_j,t_k) \ge 0$, on $\Omega^{-M,N}$, $L_2^{M,N}\Psi(x_j,t_k) \ge 0$ on $\Omega^{+M,N}$ and $D_x^+\Psi(x_{N/2},t_k) - D_x^-\Psi(x_{N/2},t_k) \le 0$ imply that $\Psi(x_j,t_k) \ge 0$ on $\overline{\Omega}^{M,N}$.*

Proof Let j^*, k^* be such that $\Psi(x_{j^*},t_{k^*}) = \min_{\overline{\Omega}^{M,N}} \Psi(x_{j^*},t_{k^*})$ and assume that the lemma is false. Then $\Psi(x_{j^*},t_{k^*}) < 0$. From the hypotheses we have $(x_{j^*},t_{k^*}) \notin \Gamma^{M,N}$, $\Psi(x_{j^*},t_{k^*}) - \Psi(x_{j^*},t_{k^*-1}) \le 0, \Psi(x_{j^*},t_{k^*}) - \Psi(x_{j^*-1},t_{k^*}) \le 0, \Psi(x_{j^*+1}, t_{k^*}) - \Psi(x_{j^*},t_{k^*}) \ge 0$ so $D_t^-\psi(x_{j^*},t_{k^*}) \le 0, \delta^2\Psi(x_{j^*},t_{k^*}) \ge 0$. It follows that

$$L_1^{M,N}\Psi(x_{j^*},t_{k^*}) = D_t^-\psi(x_{j^*},t_{k^*}) - \varepsilon\delta^2\Psi(x_{j^*},t_{k^*}) + a(x_{j^*},t_{k^*})\Psi(x_{j^*}) < 0,$$

which is a contradiction. If $(x_{j^*},t_{k^*}) \in \Omega^{+M,N}$, a similar argument shows that

$$\begin{aligned} L_2^{M,N}\Psi(x_{j^*},t_{k^*}) = D_t^-\psi(x_{j^*},t_{k^*}) - \varepsilon\delta^2\Psi(x_{j^*},t_{k^*}) + a(x_{j^*},t_{k^*})\Psi(x_{j^*},t_{k^*}) \\ +b(x_{j^*},t_{k^*})\Psi(x_{j^*}-1,t_{k^*}) < 0, \end{aligned}$$

which is a contradiction. Finally if $x_{j^*} = x_{N/2}$, then

$$D_x^-\Psi(x_{N/2},t_{k^*}) \le 0 \le D_x^+\Psi(x_{N/2},t_{k^*}) \le D_x^-\Psi(x_{N/2},t_{k^*}), \text{ by the hypothesis}$$

and so

$$\Psi(x_{\frac{N}{2}-1},t_{k^*}) = \Psi(x_{N/2},t_{k^*}) = \Psi(x_{\frac{N}{2}+1},t_{k^*}) < 0.$$

Then $L_1^N\Psi(x_{\frac{N}{2}-1},t_{k^*}) < 0$, a contradiction. This concludes the proof of the lemma.

An immediate consequence of this is the following discrete stability result.

Lemma 8 *For any mesh function Ψ on $\overline{\Omega}^{M,N}$,*

$$|\Psi(x_j,t_k)| \le \max\left\{||\Psi||_{\Gamma^{M,N}}, \frac{1}{\alpha}||L_1^{M,N}\Psi||_{\Omega^{-M,N}}, \frac{1}{\alpha}||L_2^{M,N}\Psi||_{\Omega^{+M,N}}\right\},$$

$0 \le j \le N, 0 \le k \le M.$

7 Error Estimate

Analogous to the continuous case, the discrete solution U can be decomposed into V and W which are defined to be the solutions of the following discrete problems

$$\begin{aligned} L_1^{M,N}V(x_j,t_k) = g(x_j,t_k), (x_j,t_k) \in \Omega^{-M,N}, 0 \le j \le N, 0 \le k \le M \\ V(0,t_k) = v(0,t_k), V(x_{N/2-1},t_k) = v(1-,t_k), V(x_j,0) = \phi_B(x_j), \quad (20) \\ L_2^{M,N}V(x_j,t_k) = f(x_j,t_k), (x_j,t_k) \in \Omega^{+M,N}, 0 \le j \le N, 0 \le k \le M \\ V(x_{N/2+1},t_k) = v(1+,t_k), V(2,t_k) = v(2,t_k), V(x_j,0) = \phi_B(x_j) \quad (21) \end{aligned}$$

and

$$\begin{aligned} L_1^{M,N}W(x_j,t_k) = 0, (x_j,t_k) \in \Omega^{-M,N}, W(0,t_k) = w(0,t_k), 0 \le j \le N, 0 \le k \le M \\ L_2^{M,N}W(x_j,t_k) = 0, (x_j,t_k) \in \Omega^{+M,N}, W(2,t_k) = w(2,t_k), 0 \le j \le N, 0 \le k \le M \end{aligned}$$

$$V(x_{N/2+1}, t_k) + W(x_{N/2+1}, t_k) = V(x_{N/2-1}, t_k) + W(x_{N/2-1}, t_k),$$
$$D_x^- W(x_{N/2}, t_k) + D_x^- V(x_{N/2}, t_k) = D_x^+ W(x_{N/2}, t_k) + D_x^+ V(x_{N/2}, t_k).$$
$$W(x_j, 0) = 0 \tag{22}$$

The error at each point $(x_j, t_k) \in \overline{\Omega}^{M,N}$ is denoted by $e(x_j, t_k) = U(x_j, t_k) - u(x_j, t_k)$. Then the local truncation error $L^{M,N} e(x_j, t_k)$, for $j \neq N/2$, has the decomposition

$$L^{M,N} e(x_j, t_k) = L^{M,N}(V - v)(x_j, t_k) + L^{M,N}(W - w)(x_j, t_k).$$

The error in the smooth and singular components are bounded in the following theorem.

Lemma 9 *Let $v(x_j, t_k)$ denote the smooth component of the exact solution from (1) and $V(x_j, t_k)$ the smooth component of the solution from (19), then for $j \neq \frac{N}{2}$*

$$||L_1^{M,N}(V - v)(x_j, t_k)|| \leq C(M^{-1} + (N^{-1} \ln N)^2), 0 \leq j \leq \frac{N}{2} - 1, 0 \leq k \leq M, \tag{23}$$

$$||L_2^{M,N}(V - v)(x_j, t_k)|| \leq C(M^{-1} + (N^{-1} \ln N)^2), \frac{N}{2} + 1 \leq j \leq N, 0 \leq k \leq M. \tag{24}$$

Let $w(x_j, t_k)$ denote the smooth component of the exact solution from (1) and $W(x_j, t_k)$ the smooth component of the solution from (19), then for $j \neq \frac{N}{2}$

$$||L_1^{M,N}(W - w)(x_j, t_k)|| \leq C(M^{-1} + (N^{-1} \ln N)^2), 0 \leq j \leq \frac{N}{2} - 1, 0 \leq k \leq M, \tag{25}$$

$$||L_2^{M,N}(W - w)(x_j, t_k)|| \leq C(M^{-1} + (N^{-1} \ln N)^2), \frac{N}{2} + 1 \leq j \leq N, 0 \leq k \leq M. \tag{26}$$

Proof For $j \neq \frac{N}{2}$, as the expression derived for the local truncation error in V and W and estimates for the derivatives of the smooth and singular components are exactly in the form found in [2], the required bounds hold good.

At the point $x_j = x_{N/2}$,

$$(D_x^+ - D_x^-)e(x_{N/2}, t_k) = (D_x^+ - D_x^-)(U - u)(x_{N/2}, t_k), 0 \leq k \leq M$$

Recall that $(D_x^+ - D_x^-)U(x_{N/2}, t_k) = 0$. Let $h^* = h_{N/2}^- = h_{N/2}^+$, where $h_{N/2}^- = x_{N/2} - x_{N/2-1}$ and $h_{N/2}^+ = x_{N/2+1} - x_{N/2}$.
Then

$$|(D_x^+ - D_x^-)e(x_{N/2}, t_k) \leq C\frac{h^*}{\varepsilon}. \tag{27}$$

Define a set of discrete barrier functions on $\overline{\Omega}^{M,N}$ by

$$\omega(x_j, t_k) = \frac{\Pi_{l=1}^{j}(1+\sqrt{\alpha/\varepsilon}h_l)}{\Pi_{l=1}^{N/2}(1+\sqrt{\alpha/\varepsilon}h_l)}, \quad 0 \le j \le N/2 \tag{28}$$

$$\frac{\Pi_{l=j}^{N-1}(1+\sqrt{\alpha/\varepsilon}h_{l+1})}{\Pi_{l=N/2}^{N-1}(1+\sqrt{\alpha/\varepsilon}h_{l+1})}, \quad N/2 \le j \le N. \tag{29}$$

Note that

$$\omega(0, t_k) = 0, \; \omega(1, t_k) = 1, \; \omega(2, t_k) = 0 \tag{30}$$

and from (28), for $0 \le j \le N, 0 \le k \le M$,

$$0 \le \omega(x_j, t_k) \le 1. \tag{31}$$

For $(x_j, t_k) \in \overline{\Omega}^{-M,N}$

$$D_x^+\omega(x_j, t_k) = \sqrt{\alpha/\varepsilon}\omega(x_j, t_k), \tag{32}$$

$$D_x^-\omega(x_j, t_k) = \sqrt{\alpha/\varepsilon}\frac{1}{(1+\sqrt{\alpha/\varepsilon}h_j)}\omega(x_j, t_k). \tag{33}$$

and

$$\delta^2\omega(x_j, t_k) \le \frac{2\alpha}{\varepsilon}\omega(x_j, t_k). \tag{34}$$

Similarly, for $(x_j, t_k) \in \overline{\Omega}^{+M,N}$

$$D_x^+\omega(x_j, t_k) = -\sqrt{\alpha/\varepsilon}\frac{1}{(1+\sqrt{\alpha/\varepsilon}h_{j+1})}\omega(x_j, t_k),$$

$$D_x^-\omega(x_j, t_k) = -\sqrt{\alpha/\varepsilon}\omega(x_j, t_k) \quad \text{and} \quad \delta_x^2\omega(x_j, t_k) \le \frac{2\alpha}{\varepsilon}\omega(x_j, t_k). \tag{35}$$

In particular, at $x_j = x_{N/2}$, using (35), (33) and (30),

$$(D_x^+ - D_x^-)\omega(x_j, t_k) \le -\frac{C}{\sqrt{\varepsilon}}. \tag{36}$$

From (34) and (35),

$$-\varepsilon\,\delta_x^2\omega(x_j, t_k) \ge -2\alpha\omega(x_j, t_k).$$

Therefore

$$\begin{aligned} L_1^N \omega(x_j, t_k) &= D_t^- \omega(x_j, t_k) - \varepsilon\, \delta_x^2 \omega(x_j, t_k) + a(x_j, t_k)\omega(x_j, t_k) \\ &\geq 0 - 2\alpha\omega(x_j, t_k) + a(x_j, t_k)\omega(x_j, t_k) \\ &= (a(x_j, t_k) - 2\alpha)\omega(x_j, t_k) \end{aligned} \tag{37}$$

and

$$\begin{aligned} L_2^N \omega(x_j, t_k) &= D_t^- \omega(x_j, t_k) - \varepsilon\, \delta_x^2 \omega(x_j, t_k) + a(x_j, t_k)\omega(x_j, t_k) + b(x_j, t_k)\omega(x_j - 1, t_k) \\ &\geq 0 - 2\alpha\omega(x_j, t_k) + a(x_j, t_k)\omega(x_j, t_k) + b(x_j, t_k) \\ &= (a(x_j, t_k) - 2\alpha)\omega(x_j, t_k) + b(x_j, t_k). \end{aligned} \tag{38}$$

We now state and prove the main theoretical result of this paper.

Lemma 10 *Let $u(x_j, t_k)$ denote the exact solution of (1) and $U(x_j, t_k)$ the solution of (19). Then, for $0 \leq j \leq N,\ 0 \leq k \leq M$,*

$$||U(x_j, t_k) - u(x_j, t_k)|| \leq C(M^{-1} + N^{-1} \ln N). \tag{39}$$

Proof Consider the mesh function Ψ given by
$\Psi(x_j, t_k) = C_1(M^{-1} + N^{-1} \ln N) + C_2\sqrt{\alpha/\varepsilon} h^* \omega(x_j, t_k) \pm e(x_j, t_k)$, where C_1 and C_2 are constants. Then for $x_j \in \Omega_x^{-N}$,

$$L_1^N \Psi(x_j, t_k) = C_1 a(x_j, t_k)(M^{-1} + N^{-1} \ln N) + C_2\sqrt{\alpha/\varepsilon} h^* L_1^N \omega(x_j, t_k) \pm L_1^N e(x_j, t_k). \tag{40}$$

Using (37) in (40) and Lemma 9,

$$\begin{aligned} L_1^N \Psi(x_j, t_k) &\geq C_1 a(x_j, t_k)(M^{-1} + N^{-1} \ln N) \\ &+ C_2\sqrt{\alpha/\varepsilon} h^* (a(x_j, t_k) - 2\alpha)\omega(x_j, t_k) \pm C\,(M^{-1} + N^{-1} \ln N) \geq 0, \end{aligned} \tag{41}$$

for appropriate choices of C_1 and C_2. For $x_j \in \Omega_x^{+N}$,

$$\begin{aligned} L_2^N \Psi(x_j, t_k) &= C_1(a(x_j, t_k) + b(x_j, t_k))(M^{-1} + N^{-1} \ln N) \\ &+ C_2\sqrt{\alpha/\varepsilon} h^* L_2^N \omega(x_j, t_k) \pm L_2^N e(x_j, t_k). \end{aligned} \tag{42}$$

Using (38) in (42),

$$\begin{aligned} L_2^N \Psi(x_j, t_k) &\geq C_1(a(x_j, t_k) + b(x_j, t_k))(M^{-1} + N^{-1} \ln N) \\ &+ C_2\sqrt{\alpha/\varepsilon} h^* ((a(x_j, t_k) - 2\alpha)\omega(x_j, t_k) + b(x_j, t_k)) \pm C(M^{-1} + N^{-1} \ln N). \end{aligned} \tag{43}$$

Let $\lambda(x_j, t_k) = (a(x_j, t_k) - 2\alpha)\omega(x_j, t_k) + b(x_j, t_k))$. Then choosing $C_1 > \frac{C_2||\lambda||}{2\alpha} + C$, and Lemma 9, $L_2^N \Psi(x_j, t_k) \geq 0$.

Further,

$$\begin{aligned} D_x^+\Psi(1,t_k) - D_x^-\Psi(1,t_k) &\le -C_2\frac{Ch^*}{\varepsilon} \pm C\frac{h^*}{\varepsilon}, \text{ using (27) and (36)} \\ &\le 0, \text{ for proper choice of } C_2. \end{aligned} \tag{44}$$

Also, using (30), $\Psi(0,t_k) = C_1(M^{-1} + N^{-1}\ln N) \ge 0$, $\Psi(2,t_k) = C_1(M^{-1} + N^{-1}\ln N) \ge 0$, $\Psi(x_j,0) = C_1(M^{-1} + N^{-1}\ln N) \ge 0$.

Therefore, using Lemma 7 for Ψ, it follows that $\Psi(x_j,t_k) \ge 0$ for all $0 \le j \le N, 0 \le k \le M$. As, from (31), $\omega(x_j,t_k) \le 1$ for $0 \le j \le N, 0 \le k \le M$

$$|(U-u)(x_j,t_k)| \le C(M^{-1} + N^{-1}\ln N),$$

which completes the proof.

8 Numerical Illustration

The ε–uniform convergence of the numerical method proposed in this paper is illustrated through an example presented in this section. A singularly perturbed boundary value problem for a linear parabolic second order delay differential equation of reaction-diffusion type is considered for numerical illustration.

Example

$$\begin{aligned} \frac{\partial u}{\partial t}(x,t) - \varepsilon\frac{\partial^2 u}{\partial x^2}(x,t) + (2+x+t+xt+\frac{(1+x)(x)(1-x)}{6})u(x,t) - u(x-1,t) \\ = (1+x)e^t - \frac{(1+x)(x)(1-x)}{6}, \\ \text{for } (x,t) \in ((0,1)\cup(1,2)) \times [0,T], \\ u(x,t) = 1 \text{ for } x \in [-1,0]\times[0,T], u(0,t) = 1, u(x,0) = 1, u(2,t) = 1. \end{aligned} \tag{45}$$

Fixing a fine Shishkin mesh with 128 points horizontally, the problem is solved by the method suggested above. The order of convergence and the error constant are calculated for t and the results are presented in Table 1. A graph of the numerical solution is presented in the Fig. 1.

A fine uniform mesh on t with 32 points is considered. The order of convergence and the error constant are calculated for x and the results are presented in Table 2. A graph of the numerical solution is presented in the Fig. 2.

Based on the algorithm found in [5], it is to be noted that Tables 1 and 2 give the parameter-uniform order of convergence and the error constant.

Table 1 Values of D^N, p^N, p^* and $C^N_{p^*}$ for $\varepsilon = \eta/8$ and $\alpha = 0.9$

η	Number of mesh points N				
	128	256	512	1024	2048
2^{-4}	0.587E-03	0.240E-03	0.832E-04	0.253E-04	0.707E-05
2^{-6}	0.292E-03	0.836E-04	0.225E-04	0.583E-05	0.149E-05
2^{-8}	0.293E-03	0.838E-04	0.225E-04	0.585E-05	0.149E-05
2^{-10}	0.294E-03	0.840E-04	0.226E-04	0.586E-05	0.149E-05
2^{-12}	0.294E-03	0.840E-04	0.226E-04	0.586E-05	0.149E-05
2^{-14}	0.294E-03	0.840E-04	0.226E-04	0.586E-05	0.149E-05
2^{-16}	0.294E-03	0.841E-04	0.226E-04	0.586E-05	0.149E-05
2^{-18}	0.294E-03	0.841E-04	0.226E-04	0.586E-05	0.149E-05
D^N	0.587E-03	0.240E-03	0.832E-04	0.253E-04	0.707E-05
p^N	0.129E+01	0.153E+01	0.172E+01	0.184E+01	
C^N_p	0.523E+00	0.523E+00	0.444E+00	0.331E+00	0.227E+00

t-order of convergence = 0.1291767E+01

The error constant = 0.5233417E+00

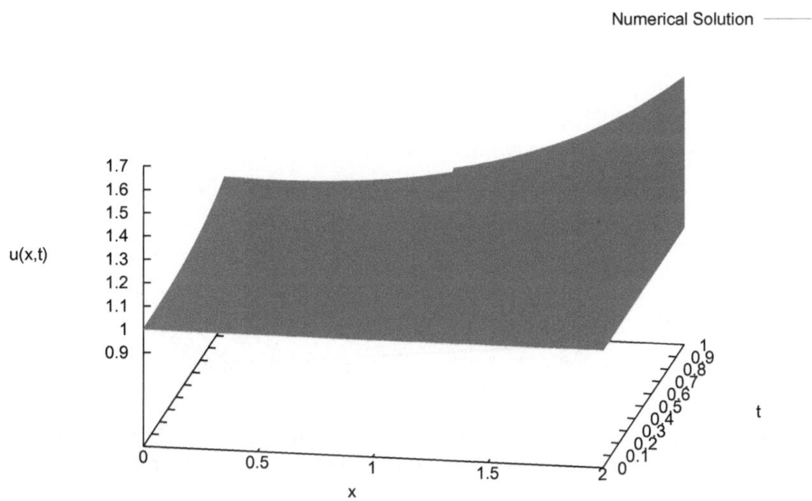

Fig. 1 The numerical solution for the problem (45), computed for $M = 1024$, $N = 128$ and $\varepsilon = 2^{-18}$. The solution $u(x, t)$ has boundary layers at $(0, t)$ and $(2, t)$ and interior layers at $(1, t)$

Table 2 Values of D^N, p^N, p^* and $C^N_{p^*}$ for $\varepsilon = \eta/16$ and $\alpha = 0.9$

η	Number of mesh points N				
	128	256	512	1024	2048
2^0	0.882E-04	0.224E-04	0.561E-05	0.140E-05	0.558E-06
2^{-3}	0.641E-03	0.174E-03	0.445E-04	0.112E-04	0.281E-05
2^{-6}	0.225E-03	0.138E-03	0.110E-03	0.587E-04	0.223E-04
2^{-9}	0.224E-03	0.138E-03	0.110E-03	0.637E-04	0.307E-04
2^{-12}	0.224E-03	0.138E-03	0.110E-03	0.637E-04	0.307E-04
D^N	0.641E-03	0.174E-03	0.110E-03	0.637E-04	0.307E-04
p^N	0.188E+01	0.662E+00	0.790E+00	0.105E+01	
C^N_p	0.432E-01	0.186E-01	0.186E-01	0.170E-01	0.129E-01

x-order of convergence = 0.6615490E+00

The error constant = 0.4316255E-01

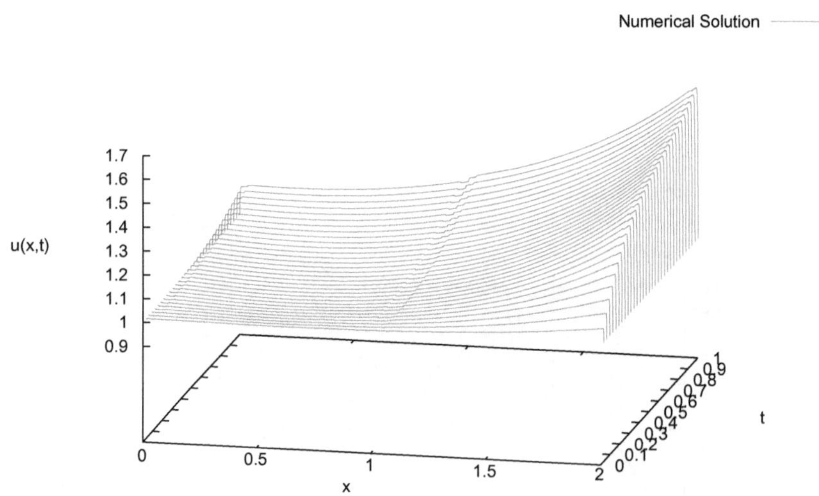

Fig. 2 The numerical solution for the problem (45), computed for $M = 32$, $N = 4096$ and $\varepsilon = 2^{-12}$. The solution $u(x, t)$ has boundary layers at $(0, t)$ and $(2, t)$ and interior layers at $(1, t)$

Acknowledgments The second and third authors acknowledge the financial support extended by the University Grants Commission, Govt. of India, through the Minor Research Project Grant—No. F MRP-5806/15 (SERO/UGC) the authors' thank the unknown refree for the valuable suggestions which resulted in this improved version of the paper.

References

1. J.J.H. Miller, E. O'Riordan, G.I. Shishkin, *Fitted Numerical Methods for Singular Perturbation Problems* (World Scientific Publishing Co., Singapore, 1996)
2. V. Franklin, M. Paramasivam, J.J.H. Miller, S. Valarmathi, Second order parameter-uniform convergence for a finite difference method for a singularly perturbed linear parabolic system. Int. J. Numer. Anal. Model. **10**(1), 178–202
3. J.J.H. Miller, E. O'Riordan, G.I. Shishkin, L.P. Shishkina, Fitted mesh methods for problems with parabolic boundary layers. Math. Proc. Royal Irish Acad. **98A**(2), 173–190 (1998)
4. E.P. Doolan, J.J.H. Miller, W.H.A. Schilders, *Uniform Numerical Methods for Problems with Initial and Boundary Layers* (Boole Press, 1980)
5. P.A. Farrell, A.F. Hegarty, J.J.H. Miller, E.O' Riordan, G.I. Shishkin, *Robust Computational Techniques for Boundary Layers* (Chapman and hall/CRC, Boca Raton, Florida,USA, 2000)
6. M. Manikandan, N. Shivaranjani, J.J.H. Miller, S. Valarmathi, A parameter uniform first order convergent numerical method for a boundary value problem for a singularly perturbed delay differential equation. Adv. Appl. Math. Springer Proc. Math. Statis. **87**, 71–88
7. A.R. Ansari, S.A. Bakr, G.I. Shishkin, A parameter-robust finite difference method for singularly perturbed delay parabolic partial differential equations. J. Comput. Appl. Math. **205**, 552–566 (2007)
8. Z. Cen, A hybrid finite difference scheme for a class of singularly perturbed delay differential equations. Neural, Parallel Scient. Comput. **16**, 303–308 (2008)
9. A. Friedman, *Partial Differential Equations of Parabolic Type* (Prentice Hall, 1964)
10. O.A. Ladyzhenskaya, N.N. Ural'tseva, *Linear and Quasilinear Elliptic Equations* (Academic Press, New York, London, 1968)

A Parameter Uniform Numerical Method for an Initial Value Problem for a System of Singularly Perturbed Delay Differential Equations with Discontinuous Source Terms

Nagarajan Shivaranjani, John J.H. Miller and Valarmathi Sigamani

Abstract In this paper an initial value problem for a system of singularly perturbed first order delay differential equations with discontinuous source terms is considered on the interval $(0, 2]$. The source terms are assumed to have simple discontinuities at the point $d \in (0, 2)$. The components of the solution exhibit initial layers and interior layers. The interior layers occuring in the solution are of two types-interior layers due to delay and interior layers due to the discontinuity of the source terms. A numerical method composed of the standard backward difference operator and a piecewise-uniform Shishkin mesh which resolves the initial and interior layers is suggested. This method is proved to be essentially first order convergent in the maximum norm uniformly in the perturbation parameters. Numerical illustrations are provided to support the theory.

Keywords Singular perturbation problems · Initial and interior layers · Delay differential equations · Discontinuous source terms · Finite difference scheme · Shishkin mesh · Parameter-uniform convergence

1 Introduction

Singularly perturbed delay differential equations play an important role in the modelling of several physical and biological phenomena like first exit time problems in modelling of activation of neuronal variability [2], bistable devices [18], evolutionary biology [3] and a variety of models for physiological processes or diseases

N. Shivaranjani · J.J.H. Miller (✉)
Bishop Heber College, Tiruchirappalli, Tamil Nadu, India
e-mail: jm@incaireland.org

N. Shivaranjani
e-mail: shivaranjaninagarajan@gmail.com

V. Sigamani
Trinity College, Dublin, Ireland
e-mail: valarmathi07@gmail.com

V. Sigamani et al. (eds.), *Differential Equations and Numerical Analysis*,
Springer Proceedings in Mathematics & Statistics 172,
DOI 10.1007/978-81-322-3598-9_8

[4–6, 13]. These systems also find applications in Belousov-Zhabotinskii reaction (BZ reaction) models and the modelling of biological oscillators [3].

In [7], the authors have considered an initial value problem for a system of singularly perturbed first order delay differential equations and have established the parameter uniform convergence of the numerical method suggested. In this paper, the following system which is similar to the one considered in [7] with discontinuous source terms is considered:

$$\mathbf{L}\mathbf{u}(x) := E\mathbf{u}'(x) + A(x)\mathbf{u}(x) + B(x)\mathbf{u}(x-1) = \mathbf{f}(x), x \in (0,d) \cup (d,2], \quad (1)$$

$$\mathbf{u}(x) = \boldsymbol{\phi}(x), x \in [-1,0], \quad (2)$$

$$\text{and } \mathbf{f}(d-) \neq \mathbf{f}(d+) \text{ for some d } \in (0,2). \quad (3)$$

For all $x \in [0,2], \mathbf{u}(x) = (u_1(x), u_2(x))^T$ and $\mathbf{f}(x) = (f_1(x), f_2(x))^T$. $E, A(x)$, $B(x)$ are 2×2 matrices. $E = diag(\boldsymbol{\varepsilon}), \boldsymbol{\varepsilon} = (\varepsilon_1, \varepsilon_2)$ with $0 < \varepsilon_1 < \varepsilon_2 \leq 1$, $B(x) = diag(\mathbf{b}(x)), \mathbf{b}(x) = (b_1(x), b_2(x))$. For all $x \in [0,2]$, it is assumed that the components $a_{ij}(x), b_i(x)$ of $A(x)$ and $B(x)$ respectively satisfy

$$b_i(x), a_{ij}(x) \leq 0 \text{ for } 1 \leq i \neq j \leq 2 \text{ and } a_{ii}(x) > \sum_{i \neq j} |a_{ij}(x) + b_i(x)| \text{ and} \quad (4)$$

$$0 < \alpha < \min_{\substack{x \in [0,2] \\ 1 \leq i \leq 2}} (\sum_{j=1}^{2} a_{ij}(x) + b_i(x)) \text{ for some } \alpha. \quad (5)$$

Further, the functions $a_{ij}(x), b_i(x), 1 \leq i, j \leq 2$ are assumed to be in $C^{(2)}([0,2])$ and $\phi_i(x)$ are assumed to be in $C^{(2)}([-1,0])$. It is to be noted that $\mathbf{L}$ can operate on functions in the domain $C^0(0,2] \cap C^1((0,d) \cup (d,2])$.

For any function $\mathbf{h}$, the jump of $\mathbf{h}$ at d is denoted by $[\mathbf{h}](d) = (\mathbf{h})(d+) - (\mathbf{h})(d-)$. Here the function $\mathbf{f}$ is assumed to have a jump of finite magnitude at d. Since $\mathbf{f}$ is discontinuous at d, the solution $\mathbf{u}$ of (1) does not necessarily have a continuous first order derivative at the point d. The cases (i) $d \in (0,1)$ and (ii) $d \in (1,2)$ are dealt with separately.

Case (i):
In this case, the components u_1 and u_2 have initial layers of width $O(\varepsilon_2)$ at $x = 0$ and interior layers of width $O(\varepsilon_2)$ at $x = 1, x = d$ and $x = 1 + d$ while the component u_1 has additional layers of width $O(\varepsilon_1)$ at $x = 0, x = 1, x = d$ and $x = 1 + d$.

Case (ii):
In this case, the components u_1 and u_2 have initial layers of width $O(\varepsilon_2)$ at $x = 0$ and interior layers of width $O(\varepsilon_2)$ at $x = 1$ and $x = d$ while the component u_1 has additional layers of width $O(\varepsilon_1)$ at $x = 0, x = 1$ and $x = d$.

In the case when $d = 1$, the solution profile is similar as that for the problem considered in [7]. The components u_1 and u_2 have initial layers of width $O(\varepsilon_2)$ at $x = 0$ and interior layers of width $O(\varepsilon_2)$ at $x = 1$ while the component u_1 has additional layers of width $O(\varepsilon_1)$ at $x = 0$ and $x = 1$. As the solution profile for this case is same as in [7], the estimates of the derivatives and the construction of Shishkin mesh for this case are as in [7]. Hence in the rest of the paper, the cases (i) and (ii) are discussed in detail.

For any vector-valued function $\mathbf{y}$ on [0, 2] the following norms are introduced: $\| \mathbf{y}(x) \| = \max_i |y_i(x)|$ and $\| \mathbf{y} \| = \sup\{\| \mathbf{y}(x) \| : x \in [0,2]\}$. For any mesh function $\mathbf{V}$ on $\overline{\Omega}^N = \{x_j\}_{j=0}^N$ the following discrete maximum norms are introduced: $\| \mathbf{V}(x_j) \| = \max_i |V_i(x_j)|$ and $\| \mathbf{V} \| = \max\{\| \mathbf{V}(x_j) \| : x_j \in \overline{\Omega}^N\}$.

Throughout this paper C denotes a generic positive constant, which is independent of x and of the two singular perturbation and discretization parameters. Furthermore, inequalities between vectors are understood in the componentwise sense.

The plan of the paper is as follows: In Sect. 2, analytical results are presented. In Sects. 3 and 4, appropriate Shishkin meshes are constructed for each case and corresponding numerical analysis is presented. In Sect. 5, the bound for the error of the discretisation is established in the maximum norm followed by numerical illustrations in Sect. 6.

2 Analytical Results

The problem (1)–(3) can be rewritten as follows for the case (i):

$$\mathbf{L}\mathbf{u}(x) := \begin{cases} \mathbf{L}_1\mathbf{u}(x) & := E\mathbf{u}'(x) + A(x)\mathbf{u}(x) = \mathbf{f}(x) - B(x)\boldsymbol{\phi}(x-1), x \in (0,d) \cup (d,1], \\ \mathbf{L}_2\mathbf{u}(x) & := E\mathbf{u}'(x) + A(x)\mathbf{u}(x) + B(x)\mathbf{u}(x-1) = \mathbf{f}(x), x \in (1,2], \\ [\varepsilon_i u_i'](d) & = [f_i](d),\ i = 1,2;\ \mathbf{u}(0) = \boldsymbol{\phi}(0) \end{cases} \tag{6}$$

and as follows for the case (ii):

$$\mathbf{L}\mathbf{u}(x) := \begin{cases} \mathbf{L}_1\mathbf{u}(x) & := E\mathbf{u}'(x) + A(x)\mathbf{u}(x) = \mathbf{f}(x) - B(x)\boldsymbol{\phi}(x-1), x \in (0,1], \\ \mathbf{L}_2\mathbf{u}(x) & := E\mathbf{u}'(x) + A(x)\mathbf{u}(x) + B(x)\mathbf{u}(x-1) = \mathbf{f}(x), x \in (0,d) \cup (d,2], \\ [\varepsilon_i u_i'](d) & = [f_i](d),\ i = 1,2;\ \mathbf{u}(0) = \boldsymbol{\phi}(0). \end{cases} \tag{7}$$

Theorem 1 *The given problem (1)–(3) has a solution* $\mathbf{u} \in \mathscr{C} = C([0,2]) \cap C^1((0,2]\backslash\{d\})$.

Proof The proof is by construction.
Case (i): Let $\mathbf{y}, \mathbf{z}, \mathbf{y}_1, \mathbf{z}_1$ be the particular solutions of

$$\begin{aligned} E\mathbf{y}'(x) + A(x)\mathbf{y}(x) &= \mathbf{f}(x) - B(x)\boldsymbol{\phi}(x-1), \quad x \in (0,d) \\ E\mathbf{z}'(x) + A(x)\mathbf{z}(x) &= \mathbf{f}(x) - B(x)\boldsymbol{\phi}(x-1), \quad x \in (d,1] \\ E\mathbf{y}_1'(x) + A(x)\mathbf{y}_1(x) &= \mathbf{f}(x) - B(x)\mathbf{y}(x-1), \quad x \in (1,1+d) \\ E\mathbf{z}_1'(x) + A(x)\mathbf{z}_1(x) &= \mathbf{f}(x) - B(x)\mathbf{z}(x-1), \quad x \in (1+d,2]. \end{aligned}$$

Consider the function,

$$\mathbf{u}(x) = \begin{cases} \mathbf{y}(x) + (\mathbf{u}(0) - \mathbf{y}(0))\boldsymbol{\psi}_1(x), & x \in [0, d) \\ \mathbf{z}(x) + \mathbf{P}\boldsymbol{\psi}_2(x), & x \in (d, 1] \\ \mathbf{y}_1(x) + \mathbf{Q}\boldsymbol{\psi}_3(x), & x \in (1, 1+d) \\ \mathbf{z}_1(x) + \mathbf{R}\boldsymbol{\psi}_4(x), & x \in (1+d, 2], \end{cases}$$

where $\boldsymbol{\psi}_1$, $\boldsymbol{\psi}_2$, $\boldsymbol{\psi}_3$, and $\boldsymbol{\psi}_4$ are solutions of

$$\begin{aligned} &E\boldsymbol{\psi}_1'(x) + A(x)\boldsymbol{\psi}_1(x) = \mathbf{0}, \quad x \in (0, d], \boldsymbol{\psi}_1(0) = \mathbf{1} \\ &E\boldsymbol{\psi}_2'(x) + A(x)\boldsymbol{\psi}_2(x) = \mathbf{0}, \quad x \in (d, 1], \boldsymbol{\psi}_2(d) = \mathbf{1} \\ &E\boldsymbol{\psi}_3'(x) + A(x)\boldsymbol{\psi}_3(x) = \mathbf{0}, \quad x \in (1, 1+d], \boldsymbol{\psi}_3(1) = \mathbf{1} \\ &E\boldsymbol{\psi}_4'(x) + A(x)\boldsymbol{\psi}_4(x) = \mathbf{0}, \quad x \in (1+d, 2], \boldsymbol{\psi}_4(1+d) = \mathbf{1} \end{aligned}$$

and $\mathbf{y}(0) = \boldsymbol{\eta}_1$, $\mathbf{z}(d) = \boldsymbol{\eta}_2$, $\mathbf{y}_1(1) = \boldsymbol{\eta}_3$, $\mathbf{z}_1(1+d) = \boldsymbol{\eta}_4$, the $\boldsymbol{\eta}_i's$ are any particular vector constants. $\mathbf{P}, \mathbf{Q}$ and $\mathbf{R}$ can be derived in the following way so as to have $\mathbf{u} \in \mathscr{C}$.

$$\begin{aligned} \mathbf{P} = \mathbf{y}(d-) + (\mathbf{u}(0) - \boldsymbol{\eta}_1)\boldsymbol{\psi}_1(d) - \mathbf{z}(d+) \\ \mathbf{Q} = \mathbf{z}(1-) + \mathbf{P} - \mathbf{y}_1(1+) \\ \mathbf{R} = \mathbf{y}_1((1+d)-) + \mathbf{Q}\boldsymbol{\psi}_3((1+d)-) - \boldsymbol{\eta}_4. \end{aligned}$$

The product between vectors is the Schur product of vectors.

Case (ii): Let $\mathbf{y}$, $\mathbf{z}$, $\mathbf{z}_1$ be the particular solutions of

$$\begin{aligned} E\mathbf{y}'(x) + A(x)\mathbf{y}(x) = \mathbf{f}(x) - B(x)\boldsymbol{\phi}(x-1), \quad x \in (0, 1] \\ E\mathbf{z}'(x) + A(x)\mathbf{z}(x) = \mathbf{f}(x) - B(x)\mathbf{y}(x-1), \quad x \in (1, d) \\ E\mathbf{z}_1'(x) + A(x)\mathbf{z}_1(x) = \mathbf{f}(x) - B(x)\mathbf{y}(x-1), \quad x \in (d, 2]. \end{aligned}$$

Consider the function,

$$\mathbf{u}(x) = \begin{cases} \mathbf{y}(x) + (\mathbf{u}(0) - \mathbf{y}(0))\boldsymbol{\psi}_1(x), & x \in [0, 1] \\ \mathbf{z}(x) + \mathbf{P}\boldsymbol{\psi}_2(x), & x \in (1, d) \\ \mathbf{z}_1(x) + \mathbf{Q}\boldsymbol{\psi}_3(x), & x \in (d, 2], \end{cases}$$

where $\boldsymbol{\psi}_1, \boldsymbol{\psi}_2, \boldsymbol{\psi}_3$ are solutions of

$$E\boldsymbol{\psi}_1'(x) + A(x)\boldsymbol{\psi}_1(x) = \mathbf{0}, \quad x \in (0,1], \boldsymbol{\psi}_1(0) = \mathbf{1}$$
$$E\boldsymbol{\psi}_2'(x) + A(x)\boldsymbol{\psi}_2(x) = \mathbf{0}, \quad x \in (1,d], \boldsymbol{\psi}_2(1) = \mathbf{1}$$
$$E\boldsymbol{\psi}_3'(x) + A(x)\boldsymbol{\psi}_3(x) = \mathbf{0}, \quad x \in (d,2], \boldsymbol{\psi}_3(d) = \mathbf{1}$$

and $\mathbf{y}(0) = \boldsymbol{\zeta}_1, \ \mathbf{z}(1) = \boldsymbol{\zeta}_2, \ \mathbf{z}_1(d) = \boldsymbol{\zeta}_3$, the $\boldsymbol{\zeta}_i'$s are any particular vector constants. $\mathbf{P}$ and $\mathbf{Q}$ can be derived in the following way so as to have $\mathbf{u} \in \mathscr{C}$.

$$\mathbf{P} = \mathbf{y}(1-) + (\mathbf{u}(0) - \boldsymbol{\zeta}_1)\boldsymbol{\psi}_1(1) - \mathbf{z}(1+)$$
$$\mathbf{Q} = \mathbf{z}(d-) + \mathbf{P}\boldsymbol{\psi}_2(d-) - \boldsymbol{\zeta}_3.$$

Analogous construction shows that the solution exists for the case when $d = 1$. In the case when $d \neq 1$, $\mathbf{u}'(1)$ exists and is continuous at 1 as $f(1)$ is well defined and is continuous at 1.

The operator $\mathbf{L}$ satisfies the following maximum principle.

Lemma 1 *Let $\boldsymbol{\psi}$ be any function in the domain of $\mathbf{L}$. Let $\boldsymbol{\psi}(0) \geq \mathbf{0}$. Then $\mathbf{L}\boldsymbol{\psi} \geq \mathbf{0}$ on $(0,d) \cup (d,2]$, $[\boldsymbol{\psi}](d) = \mathbf{0}$ implies that $\boldsymbol{\psi}(x) \geq \mathbf{0}$ on $[0,2]$.*

Proof Let $\psi_{i^*}(x^*) = \min_{i,x}\{\boldsymbol{\psi}(x)\}$ and assume $\psi_{i^*}(x^*) < 0$. Without loss of generality let $i^* = 1$. By the hypothesis, $x^* \neq 0$ and note that $\psi_1'(x^*) \leq 0$. Suppose $x^* \in (0,1] - \{d\}$, then,

$$\begin{aligned}(\mathbf{L}\boldsymbol{\psi})_1(x^*) = (\mathbf{L}_1\boldsymbol{\psi})_1(x^*) &= \varepsilon_1\psi_1'(x^*) + a_{11}(x^*)\psi_1(x^*) \\ &\quad + a_{12}(x^*)\psi_2(x^*) \\ &\leq (a_{11} + a_{12})(x^*)\psi_1(x^*) \\ &< 0,\end{aligned}$$

which is a contradiction.

Suppose $x^* \in (1,2] - \{d\}$, then,

$$\begin{aligned}(\mathbf{L}\boldsymbol{\psi})_1(x^*) = (\mathbf{L}_2\psi)_1(x^*) &= \varepsilon_1\psi_1'(x^*) + a_{11}(x^*)\psi_1(x^*) \\ &\quad + a_{12}(x^*)\psi_2(x^*) + b_1(x^*)\psi_1(x^* - 1) \\ &\leq ((a_{11} + a_{12})(x^*) + b_1(x^*))\psi_1(x^*) \\ &< 0,\end{aligned}$$

which is a contradiction.

Suppose $x^* = d$, then, $(\sum_{j=1}^{2} a_{1j}(d) + b_1(d))\psi_1(d) < 0$, and there exists a neighborhood $N_h = (d-h, d)$ such that $(\sum_{j=1}^{2} a_{1j}(x) + b_1(x))\psi_1(x) < 0$ for all $x \in N_h$. If $\psi_1'(x) < 0$ for an $x_1 \in N_h$, then $(\mathbf{L}\psi)_1(x_1) < 0$. Suppose $\psi_1'(x) > 0$ for all $x \in N_h$ then ψ_1 is an increasing function in N_h and hence cannot attain its minimum at $x = d$ which is a contradiction.

An immediate consequence of the maximum principle is the following stability result.

Lemma 2 *Let* $\mathbf{u}$ *be the solution of (1–3). Then*
$\| \mathbf{u} \| \leq C \max\{\| \mathbf{u}(0) \|, \| \mathbf{f} \|_{(0,d)\cup(d,2)}\} + C \| [\mathbf{f}](d) \|$

Proof Using the barrier function $\psi_i^{\pm}(x) = M_1 + M_2 G_i(x) \pm u_i(x)$ where $M_1 = \max\{\| \mathbf{u}(0) \|, \| \mathbf{f} \|_{(0,d)\cup(d,2)}\}$, $M_2 = \| [\mathbf{f}](d) \|$ and

$$G_i(x) = \begin{cases} \dfrac{x-d}{d} + 1, & x \in (0, d] \\ e^{-\alpha(x-d)/\varepsilon_i}, & x \in (d, 1] \\ \dfrac{(1 - exp(-\alpha(x-d)/\varepsilon_i))(x-1-d)}{d} + 1, & x \in (1, 1+d] \\ e^{-\alpha(x-(1+d))/\varepsilon_i}, & x \in (1+d, 2], \end{cases}$$

when $d \in (0, 1)$ and when $d \in (1, 2)$, $d = 1 + d_1$ for some $d_1 \in (0, 1)$ and in this case,

$$G_i(x) = \begin{cases} \dfrac{x-d_1}{d_1} + 1, & x \in (0, d_1] \\ e^{-\alpha(x-d_1)/\varepsilon_i}, & x \in (d_1, 1] \\ \dfrac{(1 - exp(-\alpha(x-d_1)/\varepsilon_i))(x-1-d_1)}{d_1} + 1, & x \in (1, d] \\ e^{-\alpha(x-d)/\varepsilon_i}, & x \in (d, 2], \end{cases}$$

and applying the maximum principle for the functions $\psi_i^{\pm}$, the result follows. Analogous proof holds for the case $d = 1$.

The solution is decomposed into smooth and singular components $\mathbf{v}$ and $\mathbf{w}$ defined to be the solutions of

$$\begin{aligned} \mathbf{L}_1\mathbf{v} &= E\mathbf{v}'(x) + A(x)\mathbf{v}(x) = \mathbf{f}(x) - B(x)\boldsymbol{\phi}(x-1) = \mathbf{g}(x) \text{ on } (0, d) \cup (d, 1] \\ \mathbf{L}_2\mathbf{v} &= E\mathbf{v}'(x) + A(x)\mathbf{v}(x) + B(x)\mathbf{v}(x-1) = \mathbf{f}(x) \text{ on } (1, 2], \end{aligned}$$

with $\mathbf{v}(0) = A^{-1}(0)(\mathbf{f}(0) - B(0)\boldsymbol{\phi}(-1))$, $\mathbf{v}(d+) = A^{-1}(d)(\mathbf{f}(d+) - B(d)\mathbf{v}(d-1))$ and
$\mathbf{L}_1\mathbf{w} = \mathbf{0}$ for $x \in (0, d) \cup (d, 1]$; $\mathbf{L}_2\mathbf{w} = \mathbf{0}$ for $x \in (1, 2]$ with $\mathbf{w}(0) = \mathbf{u}(0) - \mathbf{v}(0)$, $\mathbf{w}(d+) = \mathbf{w}(d^-) - [\mathbf{v}](d)$ respectively for case (i) and

$$\begin{aligned} \mathbf{L}_1\mathbf{v} &= E\mathbf{v}'(x) + A(x)\mathbf{v}(x) = \mathbf{f}(x) - B(x)\boldsymbol{\phi}(x-1) = \mathbf{g}(x) \text{ on } (0, 1] \\ \mathbf{L}_2\mathbf{v} &= E\mathbf{v}'(x) + A(x)\mathbf{v}(x) + B(x)\mathbf{v}(x-1) = \mathbf{f}(x) \text{ on } (1, d) \cup (d, 2], \end{aligned}$$

with $\mathbf{v}(0)=A^{-1}(0)(\mathbf{f}(0)-B(0)\boldsymbol{\phi}(-1))$, $\mathbf{v}(d+)=A^{-1}(d)(\mathbf{f}(d+)-B(d)\mathbf{v}(d-1))$ and
$\mathbf{L_1 w}=\mathbf{0}$ for $x\in(0,1]$ and $\mathbf{L_2 w}=\mathbf{0}$ for $x\in(1,d)\cup(d,2)$ with $\mathbf{w}(0)=\mathbf{u}(0)-\mathbf{v}(0)$, $\mathbf{w}(d+)=\mathbf{w}(d^-)-[\mathbf{v}](d)$ respectively for case (ii).

The bounds of the smooth component and its derivatives are contained in the following

Lemma 3 *The smooth component* $\mathbf{v}$ *satisfies the bounds*

$$||\mathbf{v}||\le C,\ \ ||\mathbf{v}'||\le C,\ \ ||v_i''||\le C\varepsilon_i^{-1}, i=1,2 \text{ on } (0,2]\backslash\{d,1+d\} \text{ for case (i)}$$

and

$$||\mathbf{v}||\le C,\ \ ||\mathbf{v}'||\le C,\ \ ||v_i''||\le C\varepsilon_i^{-1}, i=1,2 \text{ on } (0,2]\backslash\{d\} \text{ for case (ii).}$$

Proof Applying the procedure followed in [7] in the domains $[0,d)$, $(d,1]$, [1,1+d) and $(1+d,2]$ separately for the case (i) and in the domains $[0,1]$, $[1,d)$, $(d,2]$ separately for the case (ii), the result follows.

The following layer functions are defined:
$B_{p,i}(x)=e^{\frac{-\alpha(x-p)}{\varepsilon_i}}, i=1,2,\ \ p=0,1,d,1+d.$

The bounds of the singular components are derived in terms of these layer functions and are presented in the following lemma.

Lemma 4 *The singular component* $\mathbf{w}$ *satisfies the following bounds for case (i)*

$$|w_k(x)|\le\begin{cases}CB_{0,2}(x), x\in[0,d)\\ CB_{d,2}(x), x\in(d,1]\\ CB_{1,2}(x), x\in[1,1+d)\\ CB_{1+d,2}(x), x\in(1+d,2]\end{cases}\quad |w_1'(x)|\le\begin{cases}C(\varepsilon_1^{-1}B_{0,1}(x)+\varepsilon_2^{-1}B_{0,2}(x)), x\in[0,d)\\ C(\varepsilon_1^{-1}B_{d,1}(x)+\varepsilon_2^{-1}B_{d,2}(x)), x\in(d,1]\\ C(\varepsilon_1^{-1}B_{1,1}(x)+\varepsilon_2^{-1}B_{1,2}(x)), x\in[1,1+d)\\ C(\varepsilon_1^{-1}B_{1+d,1}(x)+\varepsilon_2^{-1}B_{1+d,2}(x)), x\in(1+d,2]\end{cases}$$

$$|w_2'(x)|\le\begin{cases}C\varepsilon_2^{-1}B_{0,2}(x), x\in[0,d)\\ C\varepsilon_2^{-1}B_{d,2}(x), x\in(d,1]\\ C\varepsilon_2^{-1}B_{1,2}(x), x\in[1,1+d)\\ C\varepsilon_2^{-1}B_{1+d,2}(x), x\in(1+d,2]\end{cases}\quad |w_k''(x)|\le\begin{cases}C\varepsilon_k^{-1}(\varepsilon_1^{-1}B_{0,1}(x)+\varepsilon_2^{-1}B_{0,2}(x)), x\in[0,d)\\ C\varepsilon_k^{-1}(\varepsilon_1^{-1}B_{d,1}(x)+\varepsilon_2^{-1}B_{d,2}(x)), x\in(d,1]\\ C\varepsilon_k^{-1}(\varepsilon_1^{-1}B_{1,1}(x)+\varepsilon_2^{-1}B_{1,2}(x)), x\in[1,1+d)\\ C\varepsilon_k^{-1}(\varepsilon_1^{-1}B_{1+d,1}(x)+\varepsilon_2^{-1}B_{1+d,2}(x)), x\in(1+d,2]\end{cases}$$

and for the case (ii), the following bounds are satisfied

$$|w_k(x)|\le\begin{cases}CB_{0,2}(x), x\in[0,1]\\ CB_{1,2}(x), x\in[1,d)\\ CB_{d,2}(x), x\in(d,2]\end{cases}\quad |w_1'(x)|\le\begin{cases}C(\varepsilon_1^{-1}B_{0,1}(x)+\varepsilon_2^{-1}B_{0,2}(x)), x\in[0,1]\\ C(\varepsilon_1^{-1}B_{1,1}(x)+\varepsilon_2^{-1}B_{1,2}(x)), x\in[1,d)\\ C(\varepsilon_1^{-1}B_{d,1}(x)+\varepsilon_2^{-1}B_{d,2}(x)), x\in(d,2]\end{cases}$$

$$|w_2'(x)| \le \begin{cases} C\varepsilon_2^{-1}B_{0,2}(x), x \in [0,1] \\ C\varepsilon_2^{-1}B_{1,2}(x), x \in [1,d) \\ C\varepsilon_2^{-1}B_{d,2}(x), x \in (d,2] \end{cases} \quad |w_k''(x)| \le \begin{cases} C\varepsilon_k^{-1}(\varepsilon_1^{-1}B_{0,1}(x)+\varepsilon_2^{-1}B_{0,2}(x)), x \in [0,1] \\ C\varepsilon_k^{-1}(\varepsilon_1^{-1}B_{1,1}(x)+\varepsilon_2^{-1}B_{1,2}(x)), x \in [1,d) \\ C\varepsilon_k^{-1}(\varepsilon_1^{-1}B_{d,1}(x)+\varepsilon_2^{-1}B_{d,2}(x)), x \in (d,2] \end{cases}$$

for $k = 1, 2$.

Proof Following the procedure adopted in [7], in each domain, the required bounds are derived.

There exists a unique point x^* in $(0, 1]$, such that $\varepsilon_1^{-1}B_{0,1}(x^*) = \varepsilon_2^{-1}B_{0,2}(x^*)$, $\varepsilon_1^{-1}B_{1,1}(1+x^*) = \varepsilon_2^{-1}B_{1,2}(1+x^*)$, $\varepsilon_1^{-1}B_{d,1}(d+x^*) = \varepsilon_2^{-1}B_{d,2}(d+x^*)$ and $\varepsilon_1^{-1}B_{1+d,1}(1+d+x^*) = \varepsilon_2^{-1}B_{1+d,2}(1+d+x^*)$.

The properties of this point are judicially used in deriving the novel estimates for the derivatives of the singular components.

The existence, uniqueness and properties of this point are derived in [1].

For the parameter-uniform convergence of the method suggested, the singular component is further decomposed as follows:

$$w_1(x) = w_{1,1}(x) + w_{1,2}(x), \quad w_2(x) = w_{2,1}(x) + w_{2,2}(x).$$

Following the steps in [1], it is not hard to derive the following estimates.

$$|w_{1,1}'(x)| \le C\varepsilon_1^{-1}B_{0,1}(x), \quad |w_{1,2}''(x)| \le C\varepsilon_1^{-1}\varepsilon_2^{-1}B_{0,2}(x),$$

$$|w_{2,1}'(x)| \le C\varepsilon_2^{-1}B_{0,1}(x), \quad |w_{2,2}''(x)| \le C\varepsilon_2^{-2}B_{0,2}(x), \quad x \in [0,d),$$

$$|w_{1,1}'(x)| \le C\varepsilon_1^{-1}B_{d,1}(x), \quad |w_{1,2}''(x)| \le C\varepsilon_1^{-1}\varepsilon_2^{-1}B_{d,2}(x),$$

$$|w_{2,1}'(x)| \le C\varepsilon_2^{-1}B_{d,1}(x), \quad |w_{2,2}''(x)| \le C\varepsilon_2^{-2}B_{d,2}(x), \quad x \in (d,1],$$

$$|w_{1,1}'(x)| \le C\varepsilon_1^{-1}B_{1,1}(x), \quad |w_{1,2}''(x)| \le C\varepsilon_1^{-1}\varepsilon_2^{-1}B_{1,2}(x),$$

$$|w_{2,1}'(x)| \le C\varepsilon_2^{-1}B_{1,1}(x), \quad |w_{2,2}''(x)| \le C\varepsilon_2^{-2}B_{1,2}(x), \quad x \in [1,1+d)$$

and

$$|w_{1,1}'(x)| \le C\varepsilon_1^{-1}B_{1+d,1}(x), \quad |w_{1,2}''(x)| \le C\varepsilon_1^{-1}\varepsilon_2^{-1}B_{1+d,2}(x),$$

$$|w_{2,1}'(x)| \le C\varepsilon_2^{-1}B_{1+d,1}(x), \quad |w_{2,2}''(x)| \le C\varepsilon_2^{-2}B_{1+d,2}(x), \quad x \in (1+d,2]$$

for the case (i) and

$$|w_{1,1}'(x)| \le C\varepsilon_1^{-1}B_{0,1}(x), \quad |w_{1,2}''(x)| \le C\varepsilon_1^{-1}\varepsilon_2^{-1}B_{0,2}(x),$$

$$|w_{2,1}'(x)| \le C\varepsilon_2^{-1}B_{0,1}(x), \quad |w_{2,2}''(x)| \le C\varepsilon_2^{-2}B_{0,2}(x), \quad x \in [0,1],$$

$$|w'_{1,1}(x)| \leq C\varepsilon_1^{-1}B_{1,1}(x), \quad |w''_{1,2}(x)| \leq C\varepsilon_1^{-1}\varepsilon_2^{-1}B_{1,2}(x),$$

$$|w'_{2,1}(x)| \leq C\varepsilon_2^{-1}B_{1,1}(x), \quad |w''_{2,2}(x)| \leq C\varepsilon_2^{-2}B_{1,2}(x), \quad x \in [1,d)$$

and

$$|w'_{1,1}(x)| \leq C\varepsilon_1^{-1}B_{d,1}(x), \quad |w''_{1,2}(x)| \leq C\varepsilon_1^{-1}\varepsilon_2^{-1}B_{d,2}(x),$$

$$|w'_{2,1}(x)| \leq C\varepsilon_2^{-1}B_{d,1}(x), \quad |w''_{2,2}(x)| \leq C\varepsilon_2^{-2}B_{d,2}(x), \quad x \in (d,2]$$

for the case (ii).

3 Shishkin Mesh

The Shishkin mesh $\Omega^N = \{x_j\}_{j=0}^N$ is constructed on $\overline{\Omega} = [0, 2]$ as follows for the case (i) when $\varepsilon_1 < \varepsilon_2$. In the case $\varepsilon_1 = \varepsilon_2$ a simpler construction requiring just two parameters τ and σ suffices. The interval $[0, 1]$ is subdivided into 6 sub-intervals $[0, \tau_1] \cup (\tau_1, \tau_2] \cup (\tau_2, d] \cup (d, d+\tau_3] \cup, (d+\tau_3, d+\tau_4] \cup (d+\tau_4, 1]$. The parameters $\tau_r,\ r = 1, 2, 3, 4$, which determine the points separating the uniform meshes, are defined by
$\tau_2 = \min\left\{\frac{d}{2}, \frac{\varepsilon_2}{\alpha}\ln N\right\}, \tau_1 = \min\left\{\frac{\tau_2}{2}, \frac{\varepsilon_1}{\alpha}\ln N\right\}, \tau_4 = \min\left\{\frac{1-d}{2}, \frac{\varepsilon_2}{\alpha}\ln N\right\}$ and
$\tau_3 = \min\left\{\frac{\tau_4}{2}, \frac{\varepsilon_1}{\alpha}\ln N\right\}$.

Then, on each of the sub-intervals $(\tau_2, d]$ and $(d+\tau_4, 1]$ a uniform mesh with $\frac{N}{8}$ mesh points is placed and on each of the remaining sub-intervals a uniform mesh of $\frac{N}{16}$ mesh points is placed. Similarly, the interval $[1, 2]$ is also divided into 6 sub-intervals $[1, 1+\tau_1] \cup (1+\tau_1, 1+\tau_2] \cup (1+\tau_2, 1+d] \cup (1+d, 1+d+\tau_3] \cup (1+d+\tau_3, 1+d+\tau_4] \cup (1+d+\tau_4, 2]$ having the same number of mesh intervals as in $[0, 1]$.

For case (ii), the Shishkin mesh is constructed as follows: The interval $[0, 1]$ is subdivided into 3 sub-intervals $[0, \tau_1] \cup (\tau_1, \tau_2] \cup (\tau_2, 1]$ and the interval $[1, 2]$ is divided into 6 sub-intervals $[1, 1+\tau_3] \cup (1+\tau_3, 1+\tau_4] \cup (1+\tau_4, d] \cup (d, 1+\tau_5] \cup, (1+\tau_5, 1+\tau_6] \cup (1+\tau_6, 2]$. On each of the intervals $(\tau_2, 1]$, $(1+\tau_4, d]$ and $(1+\tau_6, 2]$, a uniform mesh with $\frac{N}{6}$ points is placed and in each of the remaining intervals a uniform mesh with $\frac{N}{12}$ mesh points is placed. The parameters $\tau_r,\ r = 1, 2..., 6$ are defined as follows:

$$\tau_2 = \min\left\{\frac{1}{2}, \frac{\varepsilon_2}{\alpha}\ln N\right\}, \quad \tau_1 = \min\left\{\frac{\tau_2}{2}, \frac{\varepsilon_1}{\alpha}\ln N\right\}, \quad \tau_4 = \min\left\{\frac{d-1}{2}, \frac{\varepsilon_2}{\alpha}\ln N\right\},$$

$$\tau_3 = \min\left\{\frac{\tau_4}{2}, \frac{\varepsilon_1}{\alpha}\ln N\right\}, \tau_6 = \min\left\{\frac{2-d}{2}, \frac{\varepsilon_2}{\alpha}\ln N\right\} \text{ and } \tau_5 = \min\left\{\frac{\tau_6}{2}, \frac{\varepsilon_1}{\alpha}\ln N\right\}.$$

4 The Discrete Problem

The IVP (6)–(7) is discretised using the backward Euler scheme applied on the piecewise uniform fitted mesh $\overline{\Omega}^N$. The discrete problem for case (i) is

$$\mathbf{L}^N\mathbf{U}(x_j) = ED^-\mathbf{U}(x_j) + A(x_j)\mathbf{U}(x_j) + B(x_j)\mathbf{U}(x_j - 1) = \mathbf{f}(x_j),$$
$$j = 1(1)\frac{N}{4} - 1, \frac{N}{4} + 1(1)N, \tag{8}$$

$$\varepsilon_i(D^+ - D^-)U_i(x_{\frac{N}{4}}) = [f_i](x_{\frac{N}{4}}), i = 1, 2. \tag{9}$$

$$\mathbf{U}(0) = \mathbf{u}(0). \tag{10}$$

The discrete problem for case (ii) is

$$\mathbf{L}^N\mathbf{U}(x_j) = ED^-\mathbf{U}(x_j) + A(x_j)\mathbf{U}(x_j) + B(x_j)\mathbf{U}(x_j - 1) = \mathbf{f}(x_j),$$
$$j = 1(1)\frac{2N}{3} - 1, \frac{2N}{3} + 1(1)N, \tag{11}$$

$$\varepsilon_i(D^+ - D^-)U_i(x_{\frac{2N}{3}}) = [f_i](x_{\frac{2N}{3}}), i = 1, 2, \tag{12}$$

$$\mathbf{U}(0) = \mathbf{u}(0). \tag{13}$$

Lemma 5 *Let* $\mathbf{Z}$ *be any vector mesh function such that* $\mathbf{Z}(x_0) \geq \mathbf{0}$, $\mathbf{L}^N\mathbf{Z}(x_j) \geq \mathbf{0}$ *for all* $x_j \in \Omega^N$ *and* $(D^+ - D^-)\mathbf{Z}(x_{\frac{N}{4}}) \leq 0$ *in case (i) and* $(D^+ - D^-)\mathbf{Z}(x_{\frac{2N}{3}}) \leq \mathbf{0}$ *in case (ii) then* $\mathbf{Z}(x_j) \geq \mathbf{0}$ *for all* $x_j \in \overline{\Omega}^N$.

Proof Let x_k be such that $\mathbf{Z}$ attains its minimum on $\overline{\Omega}^N$. Further, suppose $(\mathbf{Z})_i(x_k) < 0$. If $x_k \in (0, 1]$ or if $x_k \in [1, 2]$ and $x_k \neq d$ then, $(\mathbf{L}^N\mathbf{Z})_i(x_k) < 0$ which is a contradiction. If $x_k = d$ then, $(D^-\mathbf{Z})_i(x_k) \leq 0 \leq (D^+\mathbf{Z})_i(x_k)$ as $\mathbf{Z}$ attains its minimum at x_k. Also from hypothesis, $(D^+\mathbf{Z})_i(x_k) \leq (D^-\mathbf{Z})_i(x_k)$. Hence, $(D^-\mathbf{Z})_i(x_k) \leq 0 \leq (D^+\mathbf{Z})_i(x_k) \leq (D^-\mathbf{Z})_i(x_k)$ which implies $Z_i(x_{k-1}) = Z_i(x_k) = Z_i(x_{k+1}) < 0$. Now, $(\mathbf{L}^N\mathbf{Z})_i(x_{k-1}) < 0$ which is a contradiction.

5 Error Analysis

The error at each point $x_j \in \bar{\Omega}^N$ is denoted by $\mathbf{e}(x_j) = \mathbf{U}(x_j) - \mathbf{u}(x_j)$. For case (i) when $j \neq \frac{N}{4}$, for case (ii), when $j \neq \frac{2N}{3}$ and for the case $d = 1$ when $j \neq \frac{N}{2}$, following steps as in [1], it can be derived that

$$|(\mathbf{L}^N\mathbf{e})_i(x_j)| \leq CN^{-1}\ln N,\ i = 1, 2.$$

When $j = \frac{N}{4}$ or $\frac{2N}{3}$ or $\frac{N}{2}$,

$$\begin{aligned}|(\mathbf{L}^N\mathbf{e})_i(x_j)| &\le C\varepsilon_i h^+ \max_{[x_j,x_{j+1}]} |u_i''(\eta)| \\ &\quad + C\varepsilon_i h^- \max_{[x_{j-1},x_j]} |u_i''(\theta)| \\ &\le CN^{-1}\ln N,\end{aligned}$$

where $h^+ = x_{j+1} - x_j$ and $h^- = x_j - x_{j-1}$.

Theorem 2 *Let* $\mathbf{u}$ *be the solution of the continuous problem* (6)–(7) *and* $\mathbf{U}$ *be the solution of the discrete problem* (8)–(10). *Then*

$$||\mathbf{U} - \mathbf{u}|| \le CN^{-1}\ln N. \tag{14}$$

Proof Using the barrier functions

$$\Psi_i^{\pm}(x_j) = \begin{cases} C_1(1+2x_j)N^{-1}\ln N \pm e_i(x_j), & j \le \frac{N}{4} \text{ or } \frac{2N}{3} \text{ or } \frac{N}{2} \\ C_1(d+x_j)N^{-1}\ln N \pm e_i(x_j), & j > \frac{N}{4} \text{ or } \frac{2N}{3} \text{ or } \frac{N}{2} \end{cases}$$

and the maximum principle, it is not hard to derive the required bounds.

6 Numerical Illustration

In this section, a singularly perturbed linear system of delay differential equations is considered for numerical illustration. The source term of the system has a point of discontinuity d inside the domain of definition. All the three cases when (i) $d \in (0, 1)$ (ii) $d \in (1, 2)$ and (iii) $d = 1$ are considered. Appropriate Shishkin meshes are constructed and the resulting discrete problems (8)–(13) are solved. The results are presented in the tables and figures. It is to be noted that the error constants presented in the tables are approximations to the error constant C derived in the error bound (14).

Case (i): Consider the IVP

$$\varepsilon_1 u_1'(x) + (6+x)u_1(x) - xu_2(x) - u_1(x-1) = 0.8 + 2x \tag{15}$$

$$\varepsilon_2 u_2'(x) - u_1(x) + (5+x)u_2(x) - xu_2(x-1) = 0.9 \text{ for } x \in (0, 0.4) \tag{16}$$

$$\varepsilon_1 u_1'(x) + (6+x)u_1(x) - xu_2(x) - u_1(x-1) = 2 + 2x \tag{17}$$

$$\varepsilon_2 u_2'(x) - u_1(x) + (5+x)u_2(x) - xu_2(x-1) = 3 \text{ for } x \in (0.4, 2) \tag{18}$$

$$u_1(x) = 2,\ u_2(x) = 2,\quad x \in [-1, 0]. \tag{19}$$

Table 1 Values of D^N, p^N, p^* and $C^N_{p^*}$ for $\varepsilon_1 = \eta/4$, $\varepsilon_2 = \eta$ and $\alpha = 3.9$

η	Number of mesh points N				
	512	1024	...	16384	32768
2^0	0.103E-01	0.527E-02	...	0.842E-04	0.421E-04
2^{-3}	0.442E-01	0.306E-01	...	0.102E-02	0.550E-03
2^{-6}	0.458E-01	0.305E-01	...	0.102E-02	0.549E-03
...	...	...	...	...	...
2^{-24}	0.467E-01	0.305E-01	...	0.102E-02	0.549E-03
2^{-27}	0.467E-01	0.305E-01	...	0.102E-02	0.549E-03
D^N	0.467E-01	0.306E-01	...	0.102E-02	0.550E-03
p^N	0.609E+00	0.693E+00	...	0.897E+00	
C^N_p	0.397E+01	0.397E+01	...	0.168E+01	0.138E+01
Order of convergence $=$ 0.6094098E+00					
The error constant $=$ 0.3974310E+01					

Table 2 Values of D^N, p^N, p^* and $C^N_{p^*}$ for $\varepsilon_2 = \eta$ and $\alpha = 3.9$

ε_1	Number of mesh points N				
	256	512	...	32768	65536
2^{-5}	0.306E-01	0.189E-01	...	0.189E-02	0.102E-02
2^{-6}	0.306E-01	0.189E-01	...	0.189E-02	0.102E-02
2^{-7}	0.306E-01	0.189E-01	...	0.189E-02	0.102E-02
...	...	...	...	...	...
2^{-19}	0.305E-01	0.189E-01	...	0.189E-02	0.102E-02
2^{-20}	0.305E-01	0.189E-01	...	0.189E-02	0.102E-02
D^N	0.306E-01	0.189E-01	...	0.189E-02	0.102E-02
p^N	0.693E+00	0.770E+00	...	0.886E+00	
C^N_p	0.606E+01	0.606E+01	...	0.415E+01	0.363E+01
Order of convergence $=$ 0.6934210E+00					
The error constant $=$ 0.6059335E+01					

It is to be noted that the point of discontinuity $d \in (0, 1)$. Based on the algorithm found in [19], Tables 1 and 2 gives the parameter-uniform order of convergence and the error constant. From the table it is seen that the order of convergence well agrees with the theoretical results.

The figure displays the numerical solution for the problem (15)–(19), computed for $N = 8192$, $\varepsilon_1 = 2^{-17}$, $\varepsilon_2 = 2^{-15}$ and $d = 0.4$. The components u_1 and u_2 have initial layer at $x = 0$ and interior layers at $x = d$, $x = 1$ and $x = 1 + d$ (Fig. 1).

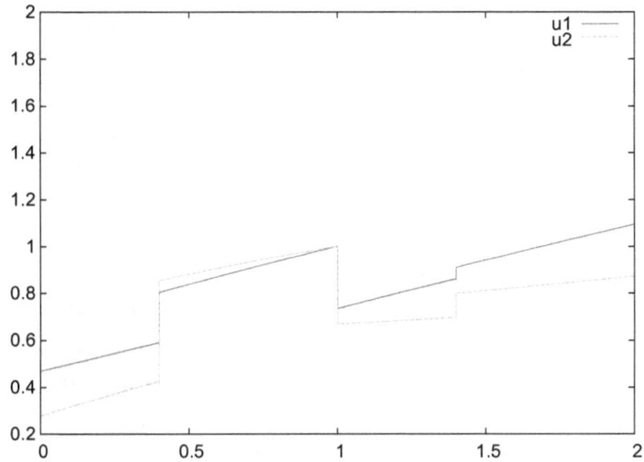

Fig. 1 Numerical solution of the IVP (15)–(19)

Case (ii): Consider the IVP

$$\varepsilon_1 u_1'(x) + (6+x)u_1(x) - xu_2(x) - u_1(x-1) = 0.8 + 2x \tag{20}$$
$$\varepsilon_2 u_2'(x) - u_1(x) + (5+x)u_2(x) - xu_2(x-1) = 0.9 \ for \ x \in (0, 1.4) \tag{21}$$
$$\varepsilon_1 u_1'(x) + (6+x)u_1(x) - xu_2(x) - u_1(x-1) = 1 + 2x \tag{22}$$
$$\varepsilon_2 u_2'(x) - u_1(x) + (5+x)u_2(x) - xu_2(x-1) = 3 \ for \ x \in (1.4, 2) \tag{23}$$
$$u_1(x) = 2 \ , u_2(x) = 2, \quad x \in [-1, 0] \tag{24}$$

It is to be noted that the point of discontinuity $d \in (1, 2)$. The maximum pointwise two mesh differences and the rate of convergence for this IVP are presented in Table 3.

This table shows that in this case also the order of convergence increases to one. The numerical solution computed for problem (20)–(24) is presented in Fig. 2 for $N = 6144$, $\varepsilon_1 = 2^{-17}$, $\varepsilon_2 = 2^{-15}$ and $d = 1.4$. In this case, the components u_1 and u_2 have initial layer at $x = 0$ and interior layers at $x = 1$ and $x = d$.

Case (iii): Consider the IVP

$$\varepsilon_1 u_1'(x) + (6+x)u_1(x) - xu_2(x) - u_1(x-1) = 0.8 + 2x \tag{25}$$
$$\varepsilon_2 u_2'(x) - u_1(x) + (5+x)u_2(x) - xu_2(x-1) = 0.9 \ for \ x \in (0, 1) \tag{26}$$
$$\varepsilon_1 u_1'(x) + (6+x)u_1(x) - xu_2(x) - u_1(x-1) = 1 + 2x \tag{27}$$
$$\varepsilon_2 u_2'(x) - u_1(x) + (5+x)u_2(x) - xu_2(x-1) = 3 \ for \ x \in (1, 2) \tag{28}$$
$$u_1(x) = 2 \ , u_2(x) = 2, \quad x \in [-1, 0] \tag{29}$$

Table 3 Values of D^N, p^N, p^* and $C^N_{p^*}$ for $\varepsilon_1 = \eta/4$, $\varepsilon_2 = \eta$ and $\alpha = 3.9$

η	Number of mesh points N				
	384	768	...	12288	24576
2^0	0.127E-01	0.654E-02	...	0.420E-03	0.210E-03
2^{-3}	0.294E-01	0.180E-01	...	0.183E-02	0.994E-03
2^{-6}	0.294E-01	0.180E-01	...	0.183E-02	0.993E-03
...	...	...	...	...	...
2^{-18}	0.293E-01	0.180E-01	...	0.183E-02	0.992E-03
2^{-21}	0.293E-01	0.180E-01	...	0.183E-02	0.992E-03
D^N	0.294E-01	0.180E-01	...	0.183E-02	0.994E-03
p^N	0.709E+00	0.758E+00	...	0.883E+00	
C^N_p	0.514E+01	0.514E+01	...	0.375E+01	0.332E+01
Order of convergence = $0.7089833E+00$					
The error constant = $0.5143387E+01$					

Fig. 2 Numerical solution of the IVP (20)–(24)

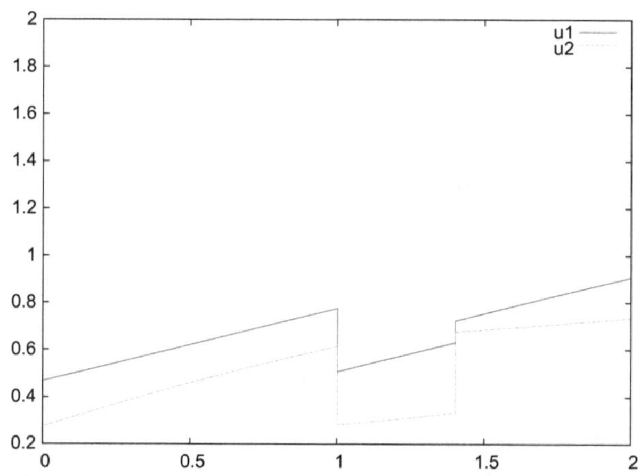

It is to be noted that the point of discontinuity $d = 1$. The maximum pointwise two mesh differences and the rate of convergence for this IVP are presented in Table 4.

This table shows that in this case also the order of convergence well agrees with the theoretical results. The numerical solution computed for problem (25)-(29) is presented in Fig. 3 for $N = 8192$, $\varepsilon_1 = 2^{-17}$, $\varepsilon_2 = 2^{-15}$ and $d = 1.0$. In this case, the components u_1 and u_2 have initial layer at $x = 0$ and interior layers at $x = 1$.

Table 4 Values of D^N, p^N, p^* and $C^N_{p^*}$ for $\varepsilon_1 = \eta/4$, $\varepsilon_2 = \eta$ and $\alpha = 3.9$

η	Number of mesh points N				
	512	1024	...	16384	32768
2^0	0.654E-02	0.332E-02	...	0.210E-03	0.105E-03
2^{-3}	0.167E-01	0.998E-02	...	0.951E-03	0.513E-03
2^{-6}	0.167E-01	0.997E-02	...	0.950E-03	0.513E-03
...	...	...	...	...	...
2^{-18}	0.167E-01	0.997E-02	...	0.950E-03	0.513E-03
2^{-21}	0.167E-01	0.997E-02	...	0.950E-03	0.513E-03
D^N	0.167E-01	0.998E-02	...	0.951E-03	0.513E-03
p^N	0.745E+00	0.809E+00	...	0.890E+00	
C^N_p	0.434E+01	0.434E+01	...	0.327E+01	0.296E+01
Order of convergence $= 0.7454905E + 00$					
The error constant $= 0.4340162E + 01$					

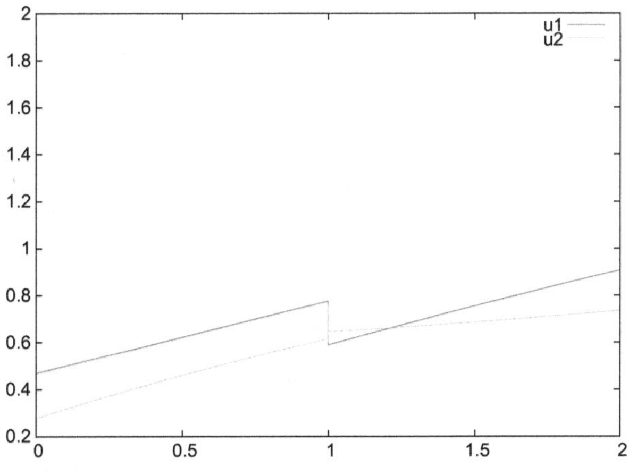

Fig. 3 Numerical solution of the IVP (25)–(29)

Acknowledgments The first author wishes to acknowledge the financial assistance extended through INSPIRE fellowship by the Department of Science and Technology, Government of India. The authors thank the unknown referee for the valuable suggestions which resulted in this improved version of the paper.

References

1. S. Valarmathi, J.J.H. Miller, A parameter uniform finite difference method for a singularly perturbed linear dynamical systems. Int. J. Numer. Anal. Model. **7**(3), 535–548 (2010)
2. C.G. Lange, R.M. Miura, Singular perturbation analysis of boundary-value problems for differential-difference equations. SIAM J. Appl. Math. **42**(3), 502–530 (1982)

3. J.D. Murray *Mathematical Biology: An Introduction*, 3rd edn. (Springer, 2002)
4. R.V. Culshaw, S. Ruan, A delay differential equation model of HIV infection of CD4+ T-Cells. Math. Biosci. **165**, 27–39 (2000)
5. A. Longtin, J.G. Milton, Complex oscillations in the human pupil light reflex with mixed and delayed feedback. Math. Biosci. **90**(1–2), 183–199 (1988)
6. P.W. Nelson, A. Perelson, Mathematical analysis of delay differential equation models of HIV - 1 infection. Math. Biosci. **179**, 73–94 (2002)
7. N. Shivaranjani, N. Ramanujam, J.J.H. Miller, S. Valarmathi, A parameter uniform method for an initial value problem for a system of singularly perturbed delay differential equations. Adv. Appl. Math. Springer Proc. Math. Statis. **87**, 127–138 (2014)
8. C.G. Lange, R.M. Miura, Singular perturbation analysis of boundary value problems for differential-difference equations. V: small shifts with layer behavior. SIAM J. Appl. Math. **54**(1), 249–272 (1994)
9. M.C. Mackey, L. Glass, Oscillations and chaos in physiological control system. Science **197**, 287–289 (1997)
10. A. Longtin, J.G. Milton, Complex oscillations in the human pupil light reflex with mixed and delayed feedback. Math. Biosci. **90**(1–2), 183–199 (1988)
11. V.Y. Glizer, Asymptotic analysis and solution of a finite-horizon H∞ control problem for singularly-perturbed linear systems with small state delay. J. Optim. Theory Appl. **117**(2), 295–325 (2003)
12. C.G. Lange, R.M. Miura, Singular perturbation analysis of boundary value problems for differential-difference equations. SIAM J. Appl. Math. **42**(3), 502–531 (1982)
13. C.G. Lange, R.M. Miura, Singular perturbation analysis of boundary value problems for differential-difference equations. VI: small shifts with rapid oscillations. SIAM J. Appl. Math. **54**(1), 273–283 (1994)
14. Z. Cen, A hybrid finite difference scheme for a class of singularly perturbed delay differential equations. Neural, Parallel Scient. Comput. **16**, 303–308 (2008)
15. T. Hongjiong, *Numerical methods for singularly perturbed delay differential equations*, An International conference on Boundary and Interior Layers, Computational and Asymptotic Methods, Toulouse, 2004
16. P. Maragatha Meenakshi, Computational techniques for solving one and two dimensional singular perturbation problems. Ph.D. Dissertation, Bharathidasan University
17. S. Nicaise, C. Xenophontos, Robust approximation of singularly perturbed delay differential equations by the hp finite element method. Comput. Meth. Appl. Math. **13**(1), 21–37 (2013)
18. M.W. Derstine, H.M. Gibbs, F.A. Hopf, D.L. Kaplan, Bifurcation gap in a hybrid optically bistable system. Phys. Rev. **26**(6), 3720–3722 (1982)
19. P.A. Farrell, A.F. Hegarty, J.J.H. Miller, E.O' Riordan, G.I. Shishkin, Robust computational techniques for boundary layers (Chapman and hall/CRC, Boca Raton, Florida,USA, 2000)

A Parameter-Uniform First Order Convergent Numerical Method for a Semi-linear System of Singularly Perturbed Second Order Delay Differential Equations

Mariappan Manikandan, John J.H. Miller and Valarmathi Sigamani

Abstract In this paper, a boundary value problem for a semi-linear system of two singularly perturbed second order delay differential equations is considered on the interval $(0, 2)$. The components of the solution of this system exhibit boundary layers at $x = 0$ and $x = 2$ and interior layers at $x = 1$. A numerical method composed of a classical finite difference operator applied on a piecewise uniform Shishkin mesh is suggested to solve the problem. The method is proved to be first order convergent in the maximum norm uniformly in the perturbation parameters. Numerical computation is described, which supports the theoretical results.

Keywords Singular perturbation problems · Boundary and interior layers · Semi-linear delay-differential equations · Finite difference scheme · Shishkin mesh · Parameter-uniform convergence

1 Introduction

Delay differential equations are common in the mathematical modelling of various physical, biological phenomena and control theory [1, 2]. A subclass of these equations consists of singularly perturbed ordinary differential equations with a delay. Such equations arise frequently in the mathematical modelling of various practical phenomena, for example, in the modelling of human pupil-light reflex [3], models of HIV infection [4], the study of bistable devices in digital electronics [5], variational problems in control theory [6], first exit time problems in modelling of activation of

M. Manikandan · V. Sigamani (✉)
Department of Mathematics, Bishop Heber College, Tiruchirappalli, Tamil Nadu, India
e-mail: valarmathi07@gmail.com

M. Manikandan
e-mail: manimaths89@yahoo.com

J.H.H. Miller
Trinity College, Dublin, Ireland
e-mail: jm@incaireland.org

V. Sigamani et al. (eds.), *Differential Equations and Numerical Analysis*,
Springer Proceedings in Mathematics & Statistics 172,
DOI 10.1007/978-81-322-3598-9_9

neuronal variability [7], evolutionary biology [8], mathematical ecology [9], population dynamics [10] and in a variety of models for physiological processes [11].

Investigation of boundary value problems for singularly perturbed linear second-order differential-difference equations was initiated by Lange and Miura [7].

The singularly perturbed boundary value problem for a semi-linear system of delay differential equations under consideration here is

$$\mathbf{T}\mathbf{u}(x) := \; -E\,\mathbf{u}''(x) + \mathbf{f}(x,\,\mathbf{u}) + B(x)\,\mathbf{u}(x-1) = \mathbf{0} \text{ on } \Omega = (0,2) \tag{1}$$

$$\text{with } \mathbf{u} = \boldsymbol{\phi} \text{ on } [-1,0] \text{ and } \mathbf{u}(2) = \mathbf{l}, \tag{2}$$

where $\phi_i \in C^2([-1,0]),\ i = 1,2$. For all $x \in [0,2] = \overline{\Omega},\ \mathbf{u}(x) = (u_1(x), u_2(x))^T$ and $\mathbf{f}(x,\,\mathbf{u}) = (f_1(x,\,\mathbf{u}), f_2(x,\,\mathbf{u}))^T$. E, and $B(x)$ are 2×2 matrices. $E = \text{diag}(\boldsymbol{\varepsilon})$, $\boldsymbol{\varepsilon} = (\varepsilon_1, \varepsilon_2)$ with $0 < \varepsilon_1 < \varepsilon_2 \leq 1,\ B(x)$=diag$(\mathbf{b}(x)),\ \mathbf{b}(x) = (b_1(x), b_2(x))$. The special cases $\varepsilon_2 = 1$ and $\varepsilon_1 = \varepsilon_2$ are simpler and could be treated with numerical schemes on modified meshes. For all $(x,\,\mathbf{u}) \in \overline{\Omega} \times R^2$, it is assumed that the nonlinear terms satisfy

$$\frac{\partial f_k(x,\,\mathbf{u})}{\partial u_k} \geq \beta > 0, \quad \frac{\partial f_k(x,\,\mathbf{u})}{\partial u_j} \leq 0,\ k, j = 1,2,\ k \neq j, \tag{3}$$

$$\min_{\substack{x\in[0,2]\\ i=1,2}} \left(\sum_{j=1}^{2} \frac{\partial f_i(x,\,\mathbf{u})}{\partial u_j} + b_i(x) \right) \geq \alpha > 0, \text{ for some } \alpha, \tag{4}$$

$$b_i(x) \leq 0,\ i = 1,2. \tag{5}$$

Further, it is assumed that $f_i \in C^2(\overline{\Omega} \times R^2)$ and $b_i \in C^2(\overline{\Omega}),\ i = 1,2$. The above assumptions ensure that $\mathbf{u} \in \mathscr{C}$ where $\mathscr{C} = C^0(\overline{\Omega}) \cap C^1(\Omega) \cap C^2((0,1) \cup (1,2))$.

The components u_1 and u_2 have boundary layers of width $O(\varepsilon_2)$ at $x = 0$ and $x = 2$ and interior layers of width $O(\varepsilon_2)$ at $x = 1$, while the component u_1 has additional boundary layers of width $O(\varepsilon_1)$ at $x = 0$ and $x = 2$ and interior layers of width $O(\varepsilon_1)$ at $x = 1$.

The problem (1)–(2) can be rewritten as

$$-\,E\,\mathbf{u}''(x) + \mathbf{f}(x,\,\mathbf{u}) + B(x)\,\boldsymbol{\phi}(x-1) = \mathbf{0} \text{ on } (0,1), \tag{6}$$

$$-\,E\,\mathbf{u}''(x) + \mathbf{f}(x,\,\mathbf{u}) + B(x)\,\mathbf{u}(x-1) = \mathbf{0} \text{ on } (1,2), \tag{7}$$

$$\mathbf{u}(0) = \boldsymbol{\phi}(0),\quad \mathbf{u}(2) = \mathbf{l}, \mathbf{u}(1-) = \mathbf{u}(1+) \text{ and } \mathbf{u}'(1-) = \mathbf{u}'(1+). \tag{8}$$

or more concisely as

$$\mathbf{T}_1\mathbf{u}(x) := -E\ \mathbf{u}''(x) + \mathbf{g}(x,\ \mathbf{u}) = \mathbf{0} \text{ on } (0, 1), \tag{9}$$

$$\mathbf{T}_2\mathbf{u}(x) := -E\ \mathbf{u}''(x) + \mathbf{f}(x,\ \mathbf{u}) + B(x)\ \mathbf{u}(x-1) = \mathbf{0} \text{ on } (1, 2), \tag{10}$$

where $\mathbf{g}(x,\ \mathbf{u}) = \mathbf{f}(x,\ \mathbf{u}) + B(x)\ \boldsymbol{\phi}(x-1)$.

The reduced problem corresponding to (9)–(10) is defined by

$$\mathbf{g}(x,\ \mathbf{r}) = \mathbf{0} \text{ on } (0, 1), \tag{11}$$

$$\mathbf{f}(x,\ \mathbf{r}) + B(x)\ \mathbf{r}(x-1) = \mathbf{0} \text{ on } (1, 2). \tag{12}$$

The implicit function theorem and conditions (3)–(5) ensure the existence of a unique solution for (11) and (12).

The solution $\mathbf{r}$ has derivatives which are bounded independently of ε_1 and ε_2.

Hence,

$$|r_1^{(k)}(x)| \le C, \quad |r_2^{(k)}(x)| \le C, \quad k = 0, 1, 2, 3, 4, \quad x \in [0, 2]. \tag{13}$$

For any vector-valued function $\mathbf{y}$ on $[0, 2]$ the following norms are introduced:

$\parallel \mathbf{y}(x) \parallel = \max_i |y_i(x)|$ and $\parallel \mathbf{y} \parallel = \sup\{\parallel \mathbf{y}(x) \parallel : x \in [0, 2]\}$. For any mesh function $\mathbf{V}$ on $\overline{\Omega}^N = \{x_j\}_{j=0}^N$ the following discrete maximum norms are introduced: $\parallel \mathbf{V}(x_j) \parallel = \max_i |V_i(x_j)|$ and $\parallel \mathbf{V} \parallel = \max\{\parallel \mathbf{V}(x_j) \parallel : x_j \in \overline{\Omega}^N\}$.

Throughout the paper C denotes a generic positive constant, which is independent of x and of all singular perturbation and discretization parameters. Furthermore, inequalities between vectors are understood in the componentwise sense.

2 Analytical Results

The following Shishkin decomposition of the solution $\mathbf{u}$ of (1)–(2) is considered:

$$\mathbf{u} = \mathbf{v} + \mathbf{w}$$

where the smooth component $\mathbf{v}(x)$ is the solution of

$$-E\ \mathbf{v}''(x) + \mathbf{g}(x,\ \mathbf{v}) = \mathbf{0} \text{ on } (0, 1),\ \mathbf{v}(0) = \mathbf{r}(0),\ \mathbf{v}(1-) = \mathbf{r}(1-), \tag{14}$$

$$-E\ \mathbf{v}''(x) + \mathbf{f}(x,\ \mathbf{v}) + B(x)\ \mathbf{v}(x-1) = \mathbf{0} \text{ on } (1, 2),\ \mathbf{v}(1+) = \mathbf{r}(1+),\ \mathbf{v}(2) = \mathbf{r}(2) \tag{15}$$

and the singular component $\mathbf{w}(x)$ is the solution of

$$-E\ \mathbf{w}''(x) + \mathbf{g}(x,\ \mathbf{v}+\mathbf{w}) - \mathbf{g}(x,\ \mathbf{v}) = \mathbf{0} \text{ on } (0,1), \tag{16}$$

$$-E\ \mathbf{w}''(x) + \mathbf{f}(x,\ \mathbf{v}+\mathbf{w}) - \mathbf{f}(x,\ \mathbf{v}) + B(x)\ \mathbf{w}(x-1) = \mathbf{0} \text{ on } (1,2), \tag{17}$$

$$\mathbf{w}(0) = \mathbf{u}(0) - \mathbf{v}(0),\ \mathbf{w}(2) = \mathbf{u}(2) - \mathbf{v}(2),\ [\mathbf{w}](1) = -[\mathbf{v}](1),\ \text{and } [\mathbf{w}'](1) = -[\mathbf{v}'](1), \tag{18}$$

where for any function $\mathbf{h}$, the jump at x is, $[\mathbf{h}](x) = \mathbf{h}(x+) - \mathbf{h}(x-)$.

Lemma 1 *For $i = 1, 2$ and for all $x \in [0,2]$, the smooth component $\mathbf{v}(x)$ satisfies,*

$$|v_i^{(k)}(x)| \le C,\ k = 0,1,2 \quad \textit{and} \quad |v_i^{(k)}(x)| \le C\left(1 + \varepsilon_i^{1-\frac{k}{2}}\right),\ k = 3,4.$$

Proof The smooth component $\mathbf{v}$ is further decomposed as follows

$$\mathbf{v} = \tilde{\mathbf{q}} + \hat{\mathbf{q}}$$

where $\hat{\mathbf{q}}$ is the solution of

$$g_1(x,\ \hat{\mathbf{q}}) = 0, \tag{19}$$

$$-\varepsilon_2 \frac{d^2\hat{q}_2}{dx^2} + g_2(x,\ \hat{\mathbf{q}}) = 0,\quad x \in (0,1), \tag{20}$$

$$\hat{q}_2(0) = v_2(0),\ \hat{q}_1(0) = v_1(0) \tag{21}$$

and

$$f_1(x,\ \hat{\mathbf{q}}) + b_1(x)\hat{q}_1(x-1) = 0, \tag{22}$$

$$-\varepsilon_2 \frac{d^2\hat{q}_2}{dx^2} + f_2(x,\ \hat{\mathbf{q}}) + b_2(x)\hat{q}_2(x-1) = 0,\ x \in (1,2), \tag{23}$$

$$\hat{q}_2(2) = v_2(2),\ \hat{q}_1(2) = v_1(2). \tag{24}$$

On the other hand, $\tilde{\mathbf{q}}$ is the solution of

$$-\varepsilon_1 \frac{d^2\tilde{q}_1}{dx^2} + g_1(x,\ \tilde{\mathbf{q}} + \hat{\mathbf{q}}) - g_1(x,\ \hat{\mathbf{q}}) = \varepsilon_1 \frac{d^2\hat{q}_1}{dx^2}, \tag{25}$$

$$-\varepsilon_2 \frac{d^2\tilde{q}_2}{dx^2} + g_2(x,\ \tilde{\mathbf{q}} + \hat{\mathbf{q}}) - g_2(x,\ \hat{\mathbf{q}}) = 0,\ x \in (0,1), \tag{26}$$

$$\tilde{q}_1(0) = \tilde{q}_2(0) = 0,\ \tilde{q}_1(1-) = v_1(1-) - \hat{q}_1(1-) \text{ and } \tilde{q}_2(1-) = v_2(1-) - \hat{q}_2(1-) \tag{27}$$

and

$$-\varepsilon_1\frac{d^2\tilde{q}_1}{dx^2}+f_1(x,\,\tilde{\mathbf{q}}+\hat{\mathbf{q}})-f_1(x,\,\hat{\mathbf{q}})+b_1(x)\tilde{q}_1(x-1)=\varepsilon_1\frac{d^2\hat{q}_1}{dx^2}, \tag{28}$$

$$-\varepsilon_2\frac{d^2\tilde{q}_2}{dx^2}+f_2(x,\,\tilde{\mathbf{q}}+\hat{\mathbf{q}})-f_2(x,\,\hat{\mathbf{q}})+b_2(x)\tilde{q}_2(x-1)=0,\ x\in(1,2), \tag{29}$$

$$\tilde{q}_1(1+)=v_1(1+)-\hat{q}_1(1+),\ \tilde{q}_2(1+)=v_2(1+)-\hat{q}_2(1+)\text{ and }\tilde{q}_1(2)=\tilde{q}_2(2)=0. \tag{30}$$

Let $x\in[0,1]$. Using (11), (19) and (20),

$$a_{11}(x)(\hat{q}_1-r_1)+a_{12}(x)(\hat{q}_2-r_2)=0, \tag{31}$$

$$-\varepsilon_2\frac{d^2}{dx^2}(\hat{q}_2-r_2)+a_{21}(x)(\hat{q}_1-r_1)+a_{22}(x)(\hat{q}_2-r_2)=\varepsilon_2\frac{d^2r_2}{dx^2} \tag{32}$$

where $a_{ij}(x)=\dfrac{\partial g_i}{\partial u_j}(x,\,\boldsymbol{\chi}_{g_i}(x)),\ i,j=1,2$, are intermediate values.

Using (31) in (32),

$$-\varepsilon_2\frac{d^2}{dx^2}(\hat{q}_2-r_2)+\left(a_{22}(x)-\frac{a_{12}(x)a_{21}(x)}{a_{11}(x)}\right)(\hat{q}_2-r_2)=\varepsilon_2\frac{d^2r_2}{dx^2}.$$

Consider the linear operator,

$$L_1z(x):=-\varepsilon_2\,z''(x)+\left(a_{22}(x)-\frac{a_{12}(x)a_{21}(x)}{a_{11}(x)}\right)z(x)=\varepsilon_2\frac{d^2r_2}{dx^2} \tag{33}$$

where $z=\hat{q}_2-r_2$.

This operator satisfies the maximum principle in [12]. Thus, $\|\,\hat{q}_2-r_2\,\|\le C\,\varepsilon_2$ and $\|\,\dfrac{d^2(\hat{q}_2-r_2)}{dx^2}\,\|\le C$. Using the mean value theorem, $|(\hat{q}_2-r_2)'(x)|=|z'(x)|\le C\,\varepsilon_2^{\frac{1}{2}}$.

Differentiating (33) with respect to x once and twice and using the bounds of z, z' and z'', we get $|z'''(x)|\le C\left(1+\varepsilon_2^{\frac{-1}{2}}\right)$ and $|z^{(iv)}(x)|\le C\,\left(1+\varepsilon_2^{-1}\right)$.

Using the bound of z in (31), $\|\,\hat{q}_1-r_1\,\|\le\,C\,\varepsilon_2$. Hence

$$\|\,\hat{q}_2\,\|\le C,\quad \|\,\frac{d\hat{q}_2}{dx}\,\|\le C,\quad \|\,\frac{d^2\hat{q}_2}{dx^2}\,\|\le C,$$

$$\|\,\frac{d^3\hat{q}_2}{dx^3}\,\|\le C\,\varepsilon_2^{\frac{-1}{2}},\quad \|\,\frac{d^4\hat{q}_2}{dx^4}\,\|\le C\,\varepsilon_2^{-1}\ \text{ and }\ \|\,\hat{q}_1\,\|\le C.$$

Differentiating (31) with respect to x once, twice, thrice and four times and using the estimates of $\frac{d\hat{q}_2}{dx}, \frac{d^2\hat{q}_2}{dx^2}, \frac{d^3\hat{q}_2}{dx^3}$ and $\frac{d^4\hat{q}_2}{dx^4}$ and the assumption that $\varepsilon_1 < \varepsilon_2$, we get

$$\| \frac{d\hat{q}_1}{dx} \| \le C, \ \| \frac{d^2\hat{q}_1}{dx^2} \| \le C, \ \| \frac{d^3\hat{q}_1}{dx^3} \| \le C\left(1+\varepsilon_1^{\frac{-1}{2}}\right) \text{ and } \| \frac{d^4\hat{q}_1}{dx^4} \| \le C\left(1+\varepsilon_1^{-1}\right).$$

From (25), (26) and (27),

$$-\varepsilon_1 \frac{d^2\tilde{q}_1}{dx^2} + a_{11}^*(x)\tilde{q}_1 + a_{12}^*(x)\tilde{q}_2 = \varepsilon_1 \frac{d^2\hat{q}_1}{dx^2},$$

$$-\varepsilon_2 \frac{d^2\tilde{q}_2}{dx^2} + a_{21}^*(x)\tilde{q}_1 + a_{22}^*(x)\tilde{q}_2 = 0,$$

$$\tilde{q}_1(0) = \tilde{q}_2(0) = 0,\ \tilde{q}_1(1-) = v_1(1-) - \hat{q}_1(1-) \text{ and } \tilde{q}_2(1-) = v_2(1-) - \hat{q}_2(1-) \tag{34}$$

where $a_{ij}^*(x) = \frac{\partial g_i}{\partial u_j}(x,\ \boldsymbol{\eta}_{g_i}(x)),\ i, j = 1, 2$, are intermediate values.

From (34), for $i = 1, 2$,

$$\| \frac{d^k\tilde{q}_i}{dx^k} \| \le C,\ k = 0, 1, 2 \ \text{ and } \ \| \frac{d^k\tilde{q}_i}{dx^k} \| \le C\left(1+\varepsilon_i^{1-\frac{k}{2}}\right),\ k = 3, 4. \tag{35}$$

Hence from the bounds for $\tilde{\mathbf{q}}$ and $\hat{\mathbf{q}}$, the required bounds of $\mathbf{v}$ follow.

Let $x \in [1, 2]$. Using (12), (22) and (23),

$$p_{11}(x)(\hat{q}_1 - r_1) + p_{12}(x)(\hat{q}_2 - r_2) + b_1(x)(\hat{q}_1(x-1) - r_1(x-1)) = 0 \tag{36}$$

$$\begin{aligned} -\varepsilon_2 \frac{d^2}{dx^2}(\hat{q}_2 - r_2) + p_{21}(x)(\hat{q}_1 - r_1) + p_{22}(x)(\hat{q}_2 - r_2) + b_2(x)(\hat{q}_2(x-1) \\ -r_2(x-1)) = \varepsilon_2 \frac{d^2 r_2}{dx^2} \end{aligned} \tag{37}$$

where $p_{ij}(x) = \frac{\partial f_i}{\partial u_j}(x,\ \boldsymbol{\kappa}_{f_i}(x)),\ i, j = 1, 2$, are intermediate values.

Using (36) in (37),

$$\begin{aligned} -\varepsilon_2 \frac{d^2}{dx^2}(\hat{q}_2 - r_2) + \left(p_{22}(x) - \frac{p_{12}(x)p_{21}(x)}{p_{11}(x)}\right)(\hat{q}_2 - r_2) \\ -\frac{p_{21}(x)}{p_{11}(x)} b_1(x)(\hat{q}_1(x-1) - r_1(x-1)) + b_2(x)(\hat{q}_2(x-1) - r_2(x-1)) = \varepsilon_2 \frac{d^2 r_2}{dx^2}. \end{aligned}$$

Consider the linear operator,

$$
\begin{aligned}
L_2 z(x) &:= -\varepsilon_2 z''(x) + \left(p_{22}(x) - \frac{p_{12}(x)p_{21}(x)}{p_{11}(x)} \right) z(x) + b_2(x) z(x-1) \\
&= \varepsilon_2 \frac{d^2 r_2}{dx^2} + \frac{p_{21}(x)}{p_{11}(x)} b_1(x)(\hat{q}_1(x-1) - r_1(x-1))
\end{aligned} \tag{38}
$$

where $z = \hat{q}_2 - r_2$.

This operator satisfies the maximum principle in [13]. Hence using similar arguments as in the interval [0, 1] and the bounds of $\hat{\mathbf{q}}$ and $\tilde{\mathbf{q}}$ in the interval [0, 1], the required bounds in the interval [1, 2] are derived. □

From equation (16),

$$
-\varepsilon_1 w_1''(x) + s_{11}(x) w_1(x) + s_{12}(x) w_2(x) = 0, \; x \in (0, 1), \tag{39}
$$

$$
-\varepsilon_2 w_2''(x) + s_{21}(x) w_1(x) + s_{22}(x) w_2(x) = 0, \; x \in (0, 1) \tag{40}
$$

where $s_{ij}(x) = \dfrac{\partial g_i}{\partial u_j}(x, \theta_{g_i}(x)), \; i, j = 1, 2,$ are intermediate values.
And from Eq. (17),

$$
-\varepsilon_1 w_1''(x) + s_{11}^*(x) w_1(x) + s_{12}^*(x) w_2(x) + b_1(x) w_1(x-1) = 0, \; x \in (1, 2), \tag{41}
$$

$$
-\varepsilon_2 w_2''(x) + s_{21}^*(x) w_1(x) + s_{22}^*(x) w_2(x) + b_2(x) w_2(x-1) = 0, \; x \in (1, 2) \tag{42}
$$

where $s_{ij}^*(x) = \dfrac{\partial f_i}{\partial u_j}(x, \lambda_{f_i}(x)), \; i, j = 1, 2,$ are intermediate values.

The singular component is given a further decomposition

$$
\mathbf{w}(x) = \mathbf{w}^l(x) + \mathbf{w}^r(x) \tag{43}
$$

$$
\text{with } \; \mathbf{w}^l(x) = \mathbf{w}(0)\, \mathbf{w}_1^l(x) + A\, \mathbf{w}_2^l(x) \tag{44}
$$

$$
\text{satisfying } \; -E\mathbf{w}_1^{l,''}(x) + S(x)\mathbf{w}_1^l(x) = \mathbf{0}, \; x \in (0, 1) \tag{45}
$$

with $\mathbf{w}_1^l(0) = \mathbf{1}$, $\mathbf{w}_1^l(1) = \mathbf{0}$ and $\mathbf{w}_1^l(x) = \mathbf{0}$ on $(1, 2]$ where $S(x) = \begin{bmatrix} s_{11}(x) & s_{12}(x) \\ s_{21}(x) & s_{22}(x) \end{bmatrix}$,

$$
-E\mathbf{w}_2^{l,''}(x) + S^*(x)\mathbf{w}_2^l(x) + B(x)\mathbf{w}_2^l(x-1) = \mathbf{0}, \; x \in (1, 2) \tag{46}
$$

with $\mathbf{w}_2^l(1) = \mathbf{1}$, $\mathbf{w}_2^l(2) = \mathbf{0}$ and $\mathbf{w}_2^l(x) = \mathbf{0}$ on $[0, 1)$ where $S^*(x) = \begin{bmatrix} s_{11}^*(x) & s_{12}^*(x) \\ s_{21}^*(x) & s_{22}^*(x) \end{bmatrix}$ and

$$\mathbf{w}^r(x) = B\,\mathbf{w}_1^r(x) + \mathbf{w}(2)\,\mathbf{w}_2^r(x) \tag{47}$$

$$\text{satisfying} \quad -E\mathbf{w}_1^{r,\prime\prime}(x) + S(x)\mathbf{w}_1^r(x) = \mathbf{0},\ x \in (0, 1) \tag{48}$$

with $\mathbf{w}_1^r(0) = \mathbf{0}$, $\mathbf{w}_1^r(1) = \mathbf{1}$ and $\mathbf{w}_1^r(x) = \mathbf{0}$ on $(1, 2]$,

$$-E\mathbf{w}_2^{r,\prime\prime}(x) + S^*(x)\mathbf{w}_2^r(x) + B(x)\mathbf{w}_2^r(x-1) = \mathbf{0},\ x \in (1, 2) \tag{49}$$

with $\mathbf{w}_2^r(1) = \mathbf{0}$, $\mathbf{w}_2^r(2) = \mathbf{1}$ and $\mathbf{w}_2^r(x) = \mathbf{0}$ on $[0, 1)$.

Here, A and B are vector constants to be chosen in such a way that the jump conditions at $x = 1$ are satisfied.

The layer functions $B_{1,i}^l$, $B_{1,i}^r$, $B_{2,i}^l$, $B_{2,i}^r$, $B_{1,i}$, $B_{2,i}$, $i = 1, 2$, associated with the solution $\mathbf{u}$, of (1)–(2), are defined by

$$B_{1,i}^l(x) = e^{-x\sqrt{\alpha}/\sqrt{\varepsilon_i}},\ B_{1,i}^r(x) = e^{-(1-x)\sqrt{\alpha}/\sqrt{\varepsilon_i}},\ B_{1,i}(x) = B_{1,i}^l(x) + B_{1,i}^r(x),\ \text{on}\ [0, 1],$$

$$B_{2,i}^l(x) = e^{-(x-1)\sqrt{\alpha}/\sqrt{\varepsilon_i}},\ B_{2,i}^r(x) = e^{-(2-x)\sqrt{\alpha}/\sqrt{\varepsilon_i}},\ B_{2,i}(x) = B_{2,i}^l(x) + B_{2,i}^r(x),$$
on $[1, 2]$.

Lemma 2 *The singular component $\mathbf{w}(x)$ satisfies, for $i = 1, 2$ and for any $x \in [0, 1]$,*

$$|w_i^l(x)| \le C\,B_{1,2}^l(x),\quad |w_i^{l,\prime}(x)| \le C\sum_{q=i}^{2}\frac{B_{1,q}^l(x)}{\sqrt{\varepsilon_q}},\quad |w_i^{l,\prime\prime}(x)| \le C\sum_{q=i}^{2}\frac{B_{1,q}^l(x)}{\varepsilon_q},$$

$$|w_i^{l,(3)}(x)| \le C\sum_{q=1}^{2}\frac{B_{1,q}^l(x)}{\varepsilon_q^{\frac{3}{2}}},\quad |\varepsilon_i w_i^{l,(4)}(x)| \le C\sum_{q=1}^{2}\frac{B_{1,q}^l(x)}{\varepsilon_q}$$

$$\textit{and for } x \in [1, 2],\quad |w_i^l(x)| \le C\,B_{2,2}^l(x),\quad |w_i^{l,\prime}(x)| \le C\sum_{q=i}^{2}\frac{B_{2,q}^l(x)}{\sqrt{\varepsilon_q}},$$

$$|w_i^{l,\prime\prime}(x)| \le C\sum_{q=i}^{2}\frac{B_{2,q}^l(x)}{\varepsilon_q},\quad |w_i^{l,(3)}(x)| \le C\sum_{q=1}^{2}\frac{B_{2,q}^l(x)}{\varepsilon_q^{\frac{3}{2}}},\quad |\varepsilon_i w_i^{l,(4)}(x)| \le C\sum_{q=1}^{2}\frac{B_{2,q}^l(x)}{\varepsilon_q}.$$

Analogous results hold for w_i^r and its derivatives.

Proof From equations (39), (40), (41), and (42), the bounds of the singular component $\mathbf{w}$ can be derived as in [14] in the domains [0, 1] and [1, 2]. □

3 The Shishkin Mesh

A piecewise uniform Shishkin mesh with N mesh-intervals is now constructed on $[0, 2]$ as follows. Let $\Omega^N = {\Omega_1}^N \cup {\Omega_2}^N$ where ${\Omega_1}^N = \{x_j\}_{j=1}^{\frac{N}{2}-1}$, ${\Omega_2}^N = \{x_j\}_{j=\frac{N}{2}+1}^{N-1}$ and $x_{\frac{N}{2}} = 1$. Then $\overline{\Omega_1}^N = \{x_j\}_{j=0}^{\frac{N}{2}}$, $\overline{\Omega_2}^N = \{x_j\}_{j=\frac{N}{2}}^{N}$, $\overline{\Omega_1}^N \cup \overline{\Omega_2}^N = \overline{\Omega}^N = \{x_j\}_{j=0}^{N}$ and $\Gamma^N = \{0, 2\}$. As the solution exhibits overlapping layers at $x = 0$ and $x = 2$ and interior overlapping layers at $x = 1$, a Shishkin mesh is constructed to resolve these layers. The interval $[0, 1]$ is subdivided into 5 sub-intervals as follows $[0, \tau_1] \cup (\tau_1, \tau_2] \cup (\tau_2, 1 - \tau_2] \cup (1 - \tau_2, 1 - \tau_1] \cup (1 - \tau_1, 1]$. The parameter $\tau_r, r = 1, 2$, which determine the points separating the uniform meshes, are defined by

$$\tau_2 = \min\left\{\frac{1}{4}, \frac{2\sqrt{\varepsilon_2}}{\sqrt{\alpha}} \ln N\right\} \text{ and } \tau_1 = \min\left\{\frac{\tau_2}{2}, \frac{2\sqrt{\varepsilon_1}}{\sqrt{\alpha}} \ln N\right\}.$$

On the sub interval $(\tau_2, 1 - \tau_2]$ a uniform mesh with $\frac{N}{4}$ mesh points is placed and on each of the sub-intervals $[0, \tau_1]$, $(\tau_1, \tau_2]$, $(1 - \tau_2, 1 - \tau_1]$ and $(1 - \tau_1, 1]$, a uniform mesh of $\frac{N}{16}$ mesh points is placed. Similarly, the interval $(1, 2]$ is also divided into 5 sub-intervals $(1, 1 + \tau_1]$, $(1 + \tau_1, 1 + \tau_2]$, $(1 + \tau_2, 2 - \tau_2]$, $(2 - \tau_2, 2 - \tau_1]$ and $(2 - \tau_1, 2]$, using the same parameters τ_1 and τ_2. In particular, when both the parameters τ_1 and τ_2 takes on their lefthand value, the Shishkin mesh $\overline{\Omega}^N$ becomes a classical uniform mesh throughout from 0 to 2. In practice, it is convenient to take $N = 16k$, $k \geq 3$. From the above construction of $\overline{\Omega}^N$, it is clear that the transition points $\{\tau_r, 1 - \tau_r, 1 + \tau_r, 2 - \tau_r\}, r = 1, 2$, are the only points at which the mesh-size can change and that it does not necessarily change at each of these points.

4 The Discrete Problem

In this section, a classical finite difference operator with an appropriate Shishkin mesh is used to construct a numerical method for (1)–(2) which is shown later to be essentially first order parameter-uniform convergent.

The discrete two-point boundary value problem is defined to be

$$\mathbf{T}_N\mathbf{U}(x_j) := -E\,\delta^2\mathbf{U}(x_j) + \mathbf{f}(x_j, \mathbf{U}(x_j)) + B(x_j)\mathbf{U}(x_j - 1) = \mathbf{0}, \quad 1 \leq j \leq N-1, \tag{50}$$

$$\mathbf{U}(x_0) = \mathbf{u}(x_0), \; D^-\mathbf{U}(x_{N/2}) = D^+\mathbf{U}(x_{N/2}) \text{ and } \mathbf{U}(x_N) = \mathbf{u}(x_N). \tag{51}$$

The problem (50)–(51) can be rewritten as

$$\mathbf{T}_{1N}\mathbf{U}(x_j) := -E\,\delta^2\mathbf{U}(x_j) + \mathbf{g}(x_j, \mathbf{U}(x_j)) = \mathbf{0}, \quad 1 \leq j \leq \frac{N}{2} - 1, \tag{52}$$

$$\mathbf{T}_{2N}\mathbf{U}(x_j) := -E\,\delta^2\mathbf{U}(x_j) + \mathbf{f}(x_j,\, \mathbf{U}(x_j)) = -B(x_j)\mathbf{U}(x_j - 1),\quad \frac{N}{2} + 1 \le j \le N - 1, \tag{53}$$

$$\mathbf{U}(x_0) = \mathbf{u}(x_0),\ D^-\mathbf{U}(x_{N/2}) = D^+\mathbf{U}(x_{N/2}) \text{ and } \mathbf{U}(x_N) = \mathbf{u}(x_N). \tag{54}$$

For $x_j \in \Omega_1^N$,

$$(\mathbf{T}_{1N}\mathbf{Y} - \mathbf{T}_{1N}\mathbf{Z})(x_j)$$

$$= -E\,\delta^2(\mathbf{Y} - \mathbf{Z})(x_j) + \mathbf{g}(x_j,\, \mathbf{Y}(x_j)) - \mathbf{g}(x_j,\, \mathbf{Z}(x_j))$$

$$= -E\,\delta^2(\mathbf{Y} - \mathbf{Z})(x_j) + \frac{\partial \mathbf{g}}{\partial u_1}(x_j,\, \mathbf{K}(x_j))(Y_1 - Z_1) + \frac{\partial \mathbf{g}}{\partial u_2}(x_j,\, \mathbf{K}(x_j))(Y_2 - Z_2)$$

$$= \mathbf{T}'_{1N}(\mathbf{Y} - \mathbf{Z})(x_j).$$

Similarly for $x_j \in \Omega_2^N$,

$$(\mathbf{T}_{2N}\mathbf{Y} - \mathbf{T}_{2N}\mathbf{Z})(x_j)$$

$$= -E\,\delta^2(\mathbf{Y} - \mathbf{Z})(x_j) + \mathbf{f}(x_j,\, \mathbf{Y}(x_j)) - \mathbf{f}(x_j,\, \mathbf{Z}(x_j))$$

$$= -E\,\delta^2(\mathbf{Y} - \mathbf{Z})(x_j) + \frac{\partial \mathbf{f}}{\partial u_1}(x_j,\, \mathbf{M}(x_j))(Y_1 - Z_1) + \frac{\partial \mathbf{f}}{\partial u_2}(x_j,\, \mathbf{M}(x_j))(Y_2 - Z_2)$$

$$= \mathbf{T}'_{2N}(\mathbf{Y} - \mathbf{Z})(x_j)$$

where $\frac{\partial \mathbf{g}}{\partial u_i}(x_j,\, \mathbf{K}(x_j))$ and $\frac{\partial \mathbf{f}}{\partial u_i}(x_j,\, \mathbf{M}(x_j)), i = 1, 2,$ are intermediate values and $\mathbf{T}'_{1N}$ and $\mathbf{T}'_{2N}$ are the Frechet derivatives of $\mathbf{T}_{1N}$ and $\mathbf{T}_{2N}$ respectively. Since $\mathbf{T}'_{1N}$ and $\mathbf{T}'_{2N}$ are linear, they satisfy the discrete maximum principle and discrete stability result in [15]. Hence,

$$\| \mathbf{Y} - \mathbf{Z} \| \le C \| \mathbf{T}'_{1N}(\mathbf{Y} - \mathbf{Z}) \| = C \| \mathbf{T}_{1N}\mathbf{Y} - \mathbf{T}_{1N}\mathbf{Z} \| \text{ on } \Omega_1^N$$

and

$$\| \mathbf{Y} - \mathbf{Z} \| \le C \| \mathbf{T}'_{2N}(\mathbf{Y} - \mathbf{Z}) \| = C \| \mathbf{T}_{2N}\mathbf{Y} - \mathbf{T}_{2N}\mathbf{Z} \| \text{ on } \Omega_2^N.$$

$$\text{i.e. } \| \mathbf{Y} - \mathbf{Z} \| \le C \| \mathbf{T}_{1N}\mathbf{Y} - \mathbf{T}_{1N}\mathbf{Z} \| \text{ on } \Omega_1^N \tag{55}$$

$$\text{and } \| \mathbf{Y} - \mathbf{Z} \| \le C \| \mathbf{T}_{2N}\mathbf{Y} - \mathbf{T}_{2N}\mathbf{Z} \| \text{ on } \Omega_2^N. \tag{56}$$

Lemma 3 *Let* $\mathbf{u}$ *be the solution of the problem* (1)–(2) *and* $\mathbf{U}$ *be the solution of the discrete problem* (52)–(54). *Then for* $j \neq \frac{N}{2}$,

$$\parallel \mathbf{U} - \mathbf{u} \parallel \leq C\,(N^{-1} \ln N)^2. \tag{57}$$

Proof Let $x_j \in \Omega_1^N$. From (55),

$$\parallel \mathbf{U} - \mathbf{u} \parallel \leq C \parallel \mathbf{T}_{1N}\mathbf{U} - \mathbf{T}_{1N}\mathbf{u} \parallel .$$

Consider,

$$\parallel \mathbf{T}_{1N}\mathbf{u} \parallel = \parallel \mathbf{T}_{1N}\mathbf{u} - \mathbf{T}_{1N}\mathbf{U} \parallel .$$

Hence,

$$\begin{aligned} \parallel \mathbf{T}_{1N}\mathbf{u} - \mathbf{T}_{1N}\mathbf{U} \parallel &= \parallel \mathbf{T}_{1N}\mathbf{u} \parallel \\ &= \parallel \mathbf{T}_{1N}\mathbf{u} - \mathbf{T}_{1}\mathbf{u} \parallel \\ &= E \parallel (\delta^2\mathbf{u} - \mathbf{u}'')(x_j) \parallel \\ &\leq E\,(\parallel (\delta^2\mathbf{v} - \mathbf{v}'')(x_j) \parallel + \parallel (\delta^2\mathbf{w} - \mathbf{w}'')(x_j) \parallel). \end{aligned}$$

Since the bounds for $\mathbf{v}$ and $\mathbf{w}$ are the same as in [14], the required result follows on Ω_1^N.

Let $x_j \in \Omega_2^N$. From (56),

$$\begin{aligned} \parallel \mathbf{U} - \mathbf{u} \parallel &\leq C \parallel \mathbf{T}_{2N}\mathbf{U} - \mathbf{T}_{2N}\mathbf{u} \parallel \\ &\leq C \parallel B(x_j)(\mathbf{U} - \mathbf{u})(x_j - 1) \parallel, \text{ from (53)} \\ &\leq C \parallel \mathbf{U} - \mathbf{u} \parallel_{\Omega_1^N} \\ &\leq C\,(N^{-1} \ln N)^2. \end{aligned}$$

Hence for $j \neq \frac{N}{2}$, $\parallel \mathbf{U} - \mathbf{u} \parallel \leq C\,(N^{-1} \ln N)^2$. □

The error at each point $x_j \in \overline{\Omega}^N$ is denoted by $\mathbf{e}(x_j) = \mathbf{U}(x_j) - \mathbf{u}(x_j)$. At the point $x_j = x_{N/2}$, for $i = 1, 2$,

$$\begin{aligned} (D^+ - D^-)e_i(x_{\frac{N}{2}}) &= (D^+ - D^-)(U_i - u_i)(x_{\frac{N}{2}}) \\ &= (D^+ - D^-)U_i(x_{\frac{N}{2}}) - (D^+ - D^-)u_i(x_{\frac{N}{2}}). \end{aligned}$$

Recall that $(D^+ - D^-)U_i(x_{\frac{N}{2}}) = 0$. Let $h^* = \max\{h^-_{N/2}, h^+_{N/2}\}$.
Then

$$\begin{aligned}
|(D^+ - D^-)e_i(x_{\frac{N}{2}})| &= |(D^+ - D^-)u_i(x_{\frac{N}{2}})| \\
&\le |(D^+ - \frac{d}{dx})u_i(x_{\frac{N}{2}})| + |(D^- - \frac{d}{dx})u_i(x_{\frac{N}{2}})| \\
&\le \frac{1}{2}h^+_{N/2}\left|u_i''(\eta)\right|_{\eta\in(1,2)} + \frac{1}{2}h^-_{N/2}\left|u_i''(\xi)\right|_{\xi\in(0,1)} \\
&\le C\,h^* \max_{x\in(0,1)\cup(1,2)}\left|u_i''(x)\right|.
\end{aligned}$$

Therefore,

$$|(D^+ - D^-)e_i(x_{\frac{N}{2}})| \le C\,\frac{h^*}{\varepsilon_i}. \tag{58}$$

Define, for $i = 1, 2$, a set of discrete barrier functions on $\overline{\Omega}^N$ by

$$\omega_i(x_j) = \begin{cases} \dfrac{\Pi_{k=1}^{j}(1+\sqrt{\alpha}h_k/\sqrt{2\varepsilon_i})}{\Pi_{k=1}^{N/2}(1+\sqrt{\alpha}h_k/\sqrt{2\varepsilon_i})}, & 0 \le j \le N/2 \\ \\ \dfrac{\Pi_{k=j}^{N-1}(1+\sqrt{\alpha}h_{k+1}/\sqrt{2\varepsilon_i})}{\Pi_{k=N/2}^{N-1}(1+\sqrt{\alpha}h_{k+1}/\sqrt{2\varepsilon_i})}, & N/2 \le j \le N. \end{cases} \tag{59}$$

Note that

$$\omega_i(0) = 0,\;\; \omega_i(1) = 1,\;\; \omega_i(2) = 0 \tag{60}$$

and from (59), and for $0 \le j \le N/2$,

$$\omega_1(x_j) = \frac{1}{\Pi_{k=j+1}^{N/2}(1+\sqrt{\alpha}h_k/\sqrt{2\varepsilon_1})},\quad \omega_2(x_j) = \frac{1}{\Pi_{k=j+1}^{N/2}(1+\sqrt{\alpha}h_k/\sqrt{2\varepsilon_2})}.$$

From the assumption that $\varepsilon_1 < \varepsilon_2$, $\dfrac{1}{1+\sqrt{\alpha}h_k/\sqrt{2\varepsilon_1}} < \dfrac{1}{1+\sqrt{\alpha}h_k/\sqrt{2\varepsilon_2}}$ which implies that, for any $0 \le j \le N/2$,

$$\omega_1(x_j) < \omega_2(x_j). \tag{61}$$

Similarly, for any $N/2 \le j \le N$, (61) holds.
Therefore, for any $0 \le j \le N$,

$$0 \le \omega_1(x_j) < \omega_2(x_j) \le 1. \tag{62}$$

It is not hard to see that, for $i = 1, 2$,

$$\begin{aligned}(\mathbf{T}'_{1N}\boldsymbol{\omega})_i(x_j) &= -\varepsilon_i\,\delta^2\omega_i(x_j) + \sum_{l=1}^{2}\frac{\partial g_i}{\partial u_l}(x_j,\,\mathbf{K}(x_j))\omega_l(x_j)\\ &\geq -\alpha\;\omega_i(x_j) \;+\; \sum_{l=1}^{2}\frac{\partial g_i}{\partial u_l}(x_j,\,\mathbf{K}(x_j))\omega_l(x_j).\end{aligned}\tag{63}$$

And

$$\begin{aligned}(\mathbf{T}'_{2N}\boldsymbol{\omega})_i(x_j) &= -\varepsilon_i\,\delta^2\omega_i(x_j) + \sum_{l=1}^{2}\frac{\partial f_i}{\partial u_l}(x_j,\,\mathbf{M}(x_j))\omega_l(x_j)\\ &\geq -\alpha\;\omega_i(x_j) \;+\; \sum_{l=1}^{2}\frac{\partial f_i}{\partial u_l}(x_j,\,\mathbf{M}(x_j))\omega_l(x_j).\end{aligned}\tag{64}$$

We now state and prove the main theoretical result of this paper.

Theorem 1 *Let* $\mathbf{u}(x_j)$ *be the solution of the problem* (1)–(2) *and* $\mathbf{U}(x_j)$ *be the solution of the discrete problem* (50)–(51). *Then,*

$$\parallel \mathbf{U}(x_j) - \mathbf{u}(x_j) \parallel \leq C\,N^{-1}\ln N,\quad 0 \leq j \leq N.$$

Proof The result follows by using the procedure adopted in the proof of Theorem 2 in [14] to the linear operators $\mathbf{T}'_{1N}$ and $\mathbf{T}'_{2N}$. □

5 Numerical Illustration

The numerical method of applying (50)–(51) on the Shishkin mesh constructed in Sect. 3 is illustrated through an example presented in this section.

Example Consider the BVP

$$-E\;\mathbf{u}''(x) + \mathbf{f}(x,\,\mathbf{u}) + B(x)\;\mathbf{u}(x-1) = \mathbf{0} \text{ on } (0,2),$$

$$\mathbf{u}(x) \;=\; (1,\;1)^T,\;\text{ for } x \in [-1,0],\;\mathbf{u}(2) \;=\; (1,\;1)^T,$$

where $E = \text{diag}(\varepsilon_1,\,\varepsilon_2)$, $B(x) = \text{diag}(-1,\,-1)$, $\mathbf{f}(x,\,\mathbf{u}) = (u_1^2(x) - 0.01u_2(x),\, u_2^2(x))^T$.

Table 1 Values of D_{ε}^{N}, D^{N}, P^{N}, P^{*}, and C_{p*}^{N} for $\varepsilon_1 = \frac{\eta}{2}$, $\varepsilon_2 = \eta$ and $\alpha = 0.9$

η	Number of mesh points N			
	128	512	1024	2048
2^{0}	0.112E-01	0.567E-02	0.286E-02	0.144E-02
2^{-2}	0.229E-01	0.119E-01	0.607E-02	0.307E-02
2^{-4}	0.431E-01	0.232E-01	0.120E-01	0.614E-02
2^{-6}	0.762E-01	0.431E-01	0.232E-01	0.120E-01
2^{-8}	0.124E+00	0.762E-01	0.431E-01	0.232E-01
2^{-10}	0.178E+00	0.124E+00	0.762E-01	0.431E-01
2^{-12}	0.158E+00	0.132E+00	0.959E-01	0.638E-01
2^{-14}	0.158E+00	0.132E+00	0.959E-01	0.634E-01
2^{-16}	0.158E+00	0.132E+00	0.959E-01	0.634E-01
2^{-18}	0.158E+00	0.132E+00	0.959E-01	0.634E-01
D^{N}	0.178E+00	0.132E+00	0.959E-01	0.638E-01
p^{N}	0.429E+00	0.464E+00	0.588E+00	
C_{p}^{N}	0.554E+01	0.554E+01	0.541E+01	0.484E+01
Computed order of ε-uniform convergence, $p^{*} = 0.4285641$				
Computed ε-uniform error constant, $C_{p*}^{N} = 5.540243$				
The numerical method suggested is found to work very well for reasonable number of mesh points				

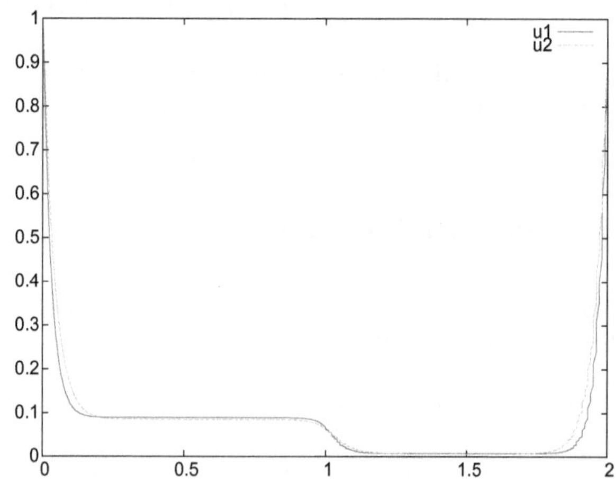

Fig. 1 Solution profile for $N = 512$ and $\eta = 2^{-8}$. The figure gives a portrait of the boundary layers and the interior layers at $x = 1$ due to the presence of the delay term for $N = 512$ and $\eta = 2^{-8}$

The maximum pointwise errors and the rate of convergence for this BVP are presented in Table 1 and the solution in Fig. 1.

References

1. N. Wang, The interior layer phenomena for a class of singularly perturbed delay-differential equations. Acta Mathematica Scientia **33B**(2), 532–542 (2013)
2. R. Jing-li, G. Wei-gao, Singularly perturbed boundary value problems for semi-linear retarted differential equations with nonlinear boundary conditions. Appl Math. Mech. **24**(12), 1450–1455 (2003)
3. A. Longtin, J.G. Milton, Complex oscillations in the human pupil light reflex with mixed and delayed feedback. Math. Biosci. **90**(1–2), 183–199 (1988)
4. R.V. Culshaw, S. Ruan, A delay-differntial equation model of HIV infection of $CD4^+$ T-cells. Math. Biosci. **165**, 27–39 (2000)
5. M.W. Derstine, H.M. Gibbs, F.A. Hopf, D.L. Kaplan, Bifurcation gap in a hybrid optically bistable system. Phys. Rev. A **26**(6), 3720–3722 (1982)
6. V.Y. Glizer, Asymptotic analysis and solution of a finite-horizon H_∞ control problem for singularly-perturbed linear systems with small state delay. J. Optim. Theory Appl. **117**, 295–325 (2003)
7. C.G. Lange, R.M. Miura, Singular perturbation analysis of boundary-value problems for differential-difference equations. VI: small shifts with rapid oscillations. SIAM J. Appl. Math. **54**(2), 273–283 (1994)
8. J.D. Murray, *Mathematical Biology: An Introduction*, 3rd edn. (Springer, 2002)
9. M. Kot, *Elements of Mathematical Ecology* (Cambridge University Press, 2001)
10. Y. Kuang, *Delay Differential Equations with Applications in Population Dynamics* (Academic Press, 1993)
11. M.C. Mackey, L. Glass, Oscillation and chaos in physiological control systems. Science **197**, 287–289 (1977)
12. J.J.H. Miller, E. O'Riordan, G.I. Shishkin, *Fitted Numerical Methods for Singular Perturbation Problems* (World Scientific Publishing Co., Singapore, 1996)
13. M. Manikandan, N. Shivaranjani, J.J.H. Miller, S. Valarmathi, A parameter uniform first order convergence numerical method for a boundary value problem for a singularly perturbed delay differential equation. Adv. Appl. Math. Springer Proc. Math. Statis. **87**, 17–88 (2014)
14. M. Manikandan, J.J.H. Miller, S. Valarmathi, *A parameter-uniform first order convergent numerical method for a system of singularly perturbed second order delay differential equations*, Proceedings of International Conference on Boundary and Interior Layers (BAIL 2014) (submitted)
15. M. Paramasivam, S. Valarmathi, J.J.H. Miller, Second order parameter-uniform convergence for a finite difference method for a singularly perturbed linear reaction diffusion system. Math. Commun. **15**(2), 587–612 (2010)

References

[illegible]